Grid-Forming Power Inverters

Grid-Forming Power Inverters: Control and Applications is the first book dedicated to addressing the operation principles, grid codes, modeling, and control of grid-forming power inverters. The book initially discusses the need for this technology due to the substantial annual integration of inverter-based renewable energy resources. The key differences between the traditional grid-following and the emerging grid-forming inverter technologies are explained. Then, the book explores in detail various topics related to grid-forming power inverters, including requirements and grid standards, modeling, control, damping power system oscillations, dynamic stability under large fault events, virtual oscillator-controlled grid-forming inverters, grid-forming inverters interfacing battery energy storage, and islanded operation of grid-forming inverters.

Features:

- Explains the key differences between grid-following and grid-forming inverters
- Explores the requirements and grid standards for grid-forming inverters
- Provides detailed modeling of virtual synchronous generators
- Explains various control strategies for grid-forming inverters
- Investigates damping of power system oscillations using grid-forming converters
- Elaborates on the dynamic stability of grid-forming inverters under large fault events
- Focuses on practical applications

Grid-Forming Power Inverters

Control and Applications

Edited by
Nabil Mohammed
Hassan Haes Alhelou
and
Behrooz Bahrani

CRC Press
Taylor & Francis Group
Boca Raton London New York

CRC Press is an imprint of the
Taylor & Francis Group, an **informa** business

First edition published 2023
by CRC Press
6000 Broken Sound Parkway NW, Suite 300, Boca Raton, FL 33487-2742

and by CRC Press
4 Park Square, Milton Park, Abingdon, Oxon, OX14 4RN

CRC Press is an imprint of Taylor & Francis Group, LLC

ISBN: 9781032298887 (hbk)
ISBN: 9781032298894 (pbk)
ISBN: 9781003302520 (ebk)

DOI: 10.1201/9781003302520

Typeset in Times
by codeMantra

Contents

Preface

The fast growth in the global electricity demand and the need to utilize environment-friendly energy resources are accelerating the advancements in power electronic systems and their implementations in modern energy systems. As a sequence, there is a substantial annual increase in the number of distributed power generations that are being installed on the demand side. In addition, clustering of small-scale generation units can form microgrids to provide reliable and optimal integration of inverter-based resources (IBRs), as they can be operated whether in grid-connected or islanded modes utilizing what so-called grid-forming inverters. In this context, grid-forming inverters are being seen as the key enabling technology for future power electronics-dominated power systems.

This book aims to provide a comprehensive overview of the grid-forming inverters. The book provides initially an overview on the differences between grid-following and grid-forming inverters. The detailed topologies, operation principles and the control techniques are elaborated. Then, the challenges in integrating of IBRs (e.g., low inertia systems and weak grids) and the requirements and grid standards for grid-forming inverters are explained. After that, the book presents the detailed modeling of the grid-forming inverters including the virtual synchronous generator (VSG). The book also covers other important topics related to grid-forming inverters, such as damping power system oscillations, dynamic stability under large fault events, virtual oscillator-controlled grid-forming inverters, grid-forming inverters interfacing battery energy storage and islanded operation of grid-forming inverters.

MATLAB® is a registered trademark of The Math Works, Inc. For product information, please contact:
The Math Works, Inc.
3 Apple Hill Drive
Natick, MA 01760-2098
Tel: 508-647-7000
Fax: 508-647-7001
E-mail: info@mathworks.com
Web: http://www.mathworks.com

Editors

Dr. Nabil Mohammed is a postdoctoral research fellow at Monash University. His Ph.D. degree was in power electronics from the Macquarie University, Sydney, NSW, Australia. In the summer of 2019, he was a visiting researcher with the Department of Energy Technology at Aalborg University, Denmark. His research interests include power electronic converters, renewable energy generation and integration in power systems, microgrids, energy storage and management systems, and modeling and control of electric systems.

Prof. Hassan Haes Alhelou is a senior member of IEEE. He is with the Department of Electrical and Computer Systems Engineering, Monash University, Clayton, VIC 3800, Australia. At the same time, he is a professor and faculty member at Tishreen University in Syria, and a consultant with Sultan Qaboos University (SQU) in Oman. Previously, he was with the School of Electrical and Electronic Engineering, University College Dublin (UCD), Dublin 4, Ireland, between 2020 and 2021, and with Isfahan University of Technology (IUT), Iran. He completed his B.Sc. from Tishreen University in 2011 and M.Sc. and Ph.D. from Isfahan University of Technology, Iran, all with honors. He was included in the 2018 and 2019 Publons and Web of Science (WoS) list of the top 1% best reviewers and researchers in the field of engineering and cross-fields over the world. He was the recipient of the Outstanding Reviewer Award from many journals, e.g., *Energy Conversion and Management (ECM)*, *ISA Transactions*, and *Applied Energy*. He was the recipient of the best young researcher in the Arab Student Forum Creative among 61 researchers from 16 countries at Alexandria University, Egypt, 2011. He also received the Excellent Paper Award 2021/2022 from IEEE *CSEE Journal of Power and Energy Systems* (SCI IF: 3.938; Q1). He has published more than 200 research papers in high-quality peer-reviewed journals and international conferences. His research papers received 2,550 citations with H-index of 26 and i-index of 56. He has authored/edited 15 books published in reputed publishers such as Springer, IET, Wiley, Elsevier, and Taylor & Francis Group. He serves as an editor in a number of prestigious journals such as *IEEE Systems Journal*, *Computers and Electrical Engineering* (CAEE-Elsevier), *IET Journal of Engineering*, and *Smart Cities*. He has also performed more than 800 reviews for high prestigious journals, including *IEEE Transactions on Power Systems*, *IEEE Transactions on Smart Grid*, *IEEE Transactions on Industrial Informatics*, *IEEE Transactions on Industrial Electronics*, *Energy Conversion and Management*, *Applied Energy*, and *International Journal of Electrical Power & Energy Systems*. He has participated in more than 15 international industrial projects over the globe. His major research interests are renewable energy systems, power systems, power system security, power system dynamics, power system cybersecurity, power system operation, control, dynamic state estimation, frequency control, smart grids, microgrids, demand response, and load shedding.

Dr. Behrooz Bahrani is a senior lecturer and the director of the Grid Innovation Hub at Monash University, Melbourne, Australia. He received a B.Sc. degree from Sharif University of Technology, Tehran, Iran; an M.Sc. degree from the University of Toronto, Toronto, Ontario, Canada; and a Ph.D. degree from the Ecole Polytechnique Federale de Lausanne (EPFL), Lausanne, Switzerland, all in electrical engineering, in 2006, 2008, and 2012, respectively. From September 2012 to September 2015, he was a postdoctoral fellow at EPFL, Purdue University, West Lafayette, IN, USA; Georgia Institute of Technology, Atlanta, GA, USA; and the Technical University of Munich, Munich, Germany. His research interests include control of power electronic converters, their applications in power systems, and grid integration of renewable energy resources.

Contributors

Hassan Haes Alhelou
Monash University
Melbourne, Australia

José Luis Rodríguez Amenedo
Department of Electrical Engineering
University Carlos III of Madrid
Leganés, Spain

Santiago Arnaltes
Department of Electrical Engineering
University Carlos III of Madrid
Leganés, Spain

Behrooz Bahrani
Department of Electrical and Computer
 Systems Engineering
Monash University
Melbourne, Australia

Jesús Castro
Electrical Engineering Department
University Carlos III of Madrid
Leganés, Spain

Mihai Ciobotaru
School of Engineering
Macquarie University
Sydney, Australia

Evangelos Farantatos
Electric Power Research Institute
Knoxville, Tennessee, USA

Damian Flynn
School of Electrical and Electronic
 Engineering
University College Dublin
Dublin, Ireland

Habib Ur Rahman Habib
Department of Electrical Engineering
Faculty of Electrical and Electronics
 Engineering
University of Engineering and
 Technology Taxila
Taxila, Pakistan and
Department of Engineering Durham
 University
Durham, England

Mehrdad Tarafdar Hagh
Faculty of Electrical and Computer
 Engineering
University of Tabriz
Tabriz, Iran

Fahim Al Hasnain
Department of Electrical and Computer
 Engineering
The University of North Carolina at Charlotte
Charlotte, North Carolina, USA

Mohammad Huque
Power Delivery & Utilization Section
Electric Power Research Institute
Knoxville, Tennessee, USA

Sukumar Kamalasadan
Department of Electrical and Computer
 Engineering
The University of North Carolina at
 Charlotte
Charlotte, North Carolina, USA

Akhlaque Ahmad Khan
Electrical Engineering Department
Integral University
Lucknow, India

Reza Deihimi Kordkandi
Faculty of Electrical and Computer
  Engineering
University of Tabriz
Tabriz, Iran

Ahmad Faiz Minai
Electrical Engineering Department
Integral University
Lucknow, India

Nabil Mohammed
Department of Electrical and Computer
  Systems Engineering
Monash University
Melbourne, Australia

Chitaranjan Phurailatpam
School of Electrical and Electronic
  Engineering
University College Dublin
Dublin, Ireland

Deepak Ramasubramanian
Power Delivery & Utilization Section
Electric Power Research Institute
Knoxville, Tennessee, USA

Sam Roozbehani
Renewable Power Generation Research
  Group
Academic Center for Education, Culture
  and Research (ACECR)
Khajeh Nasir Toosi University of
  Technology Branch
Tehran, Iran

Michael Smith
Department of Engineering Technology
  and Construction Management
The University of North Carolina at
  Charlotte
Charlotte, North Carolina, USA

Wenzong Wang
Power Delivery & Utilization Section
Electric Power Research Institute
Knoxville, Tennessee, USA

Jingzhe Xu
Department of Electrical and Computer
  Systems Engineering
Monash University
Melbourne, Australia

Mohamed Younis
Independent Electricity System
  Operator
Toronto, Ontario, Canada

Hoda Youssef
Independent Electricity System
  Operator
Toronto, Ontario, Canada

Weihua Zhou
Department of Electrical and Computer
  Systems Engineering
Monash University
Melbourne, Australia

Xianxian Zhao
School of Electrical and Electronic
  Engineering
University College Dublin
Dublin, Ireland

Introduction

This book discusses the grid-forming inverters used in the application of integration of inverter-based resources (IBRs) and describes the operation principles, the control techniques, the challenges in integrating of IBRs (e.g., low inertia systems and weak grids), modeling, and applications of grid-forming inverters. The contents of each chapter are summarized as follows:

Chapter 1: The energy sector is emphasizing the use of grid-forming inverters (GFMIs) to solve problems associated with weak grid scenarios, with recent reckon in Australia, the United Kingdom, and the United States serving as examples. Future directions are also considered in terms of voltage and frequency regulation, system strength development, and the regulatory regime. Researchers and power system engineers who are looking for solutions to the new challenges that have arisen as a result of high IBR penetration, with a focus on GFMIs, will find this chapter useful. The interactions between the IBRs and the high-power system are also discussed in this chapter. Inverters with GFMIs and traditional grid following inverters (GFLIs) have a substantial difference. In a mixed system, GFM inverters dampen frequency swings, whereas GFL inverters can exacerbate frequency problems with increased penetration.

Chapter 2: This chapter focuses on the system-level understanding of grid-forming requirements and the way forward to adopt grid-forming inverters (GFMIs) in power system networks. First, system scarcities and stability issues with high shares of inverter-based resources are discussed, followed by a comparison of the capabilities between grid-forming, grid-following, and synchronous sources to deliver the required system services. This is followed by a detailed discussion on the performance requirement of the various services that can be provided by GFMIs and the challenges associated with its large-scale implementation. Pathways for the standardized adoption of GFMIs in power systems, which can be either through grid code modifications, system services market, or utility (system-operator) owned and operated GFMI sources, are then discussed in the following section. Finally, some of the earliest proposals in grid code modification and ongoing developments to accommodate GFMIs are also covered in detail.

Chapter 3: The penetration levels of renewable resources continue to increase in power system grids around the world. Most of these resources use power electronics converters to be connected to the grid. Normal operation of power electronics converters causes decoupling from the rest of the grid and presents multiple challenges for grid operation. However, the converters can be controlled in a certain way to alleviate these issues. Grid-forming technology is one of these ways that has recently been tested. Renewable energy resources, similar to any other facility, must meet the minimum technical requirements of the grid that they are connected to. The grid requirements are essential to reliably and securely operate the electric grid and plan for the future. This chapter provides a summary of the key performance requirements for connecting power electronics converter-based resources to the public grid, focusing on grid-forming technology.

Chapter 4: As IBR percentage increases, IBRs would be expected to provide a wide variety of system services through their dynamic behavior. At present, there are many different types of emerging IBR controls being designed and developed by researchers and inverter vendors. In order to have an efficient design of these emerging control techniques, exact performance requirements must be known, which can only be specified either through standards and/or interconnection requirements from power system planners. However, writing either standards or interconnection requirements requires an idea of the capabilities of the future IBR control architecture. For a future power system, it is important that system planners do not enforce requirements on specific types of grid forming inverter control. This can result in restriction of performance and products that may enter the market place. This chapter showcases operational similarity across different control behavior that allows for the development of generic models and performance requirements.

Chapter 5: In this chapter, the modeling and control of grid-forming converters are discussed in detail. First, the basic definition and the conceptual differences between grid-forming converters and grid-following converters are discussed. Second, the model of the grid-forming converter is introduced to represent the dynamics of the control system. After that, the next section is about the control of grid-forming converter, including outer loop and inner loop. The main functionalities of the outer loop are power synchronization and voltage profile management. Also, the inner loop is developed for calculating the modulation signal for regulating the output voltage of grid-forming converter. In this section, the different control structures such as cascaded control and direct voltage control with linear and fractional order sliding mode controllers are discussed. Additionally, the design of controllers is demonstrated and verified with case studies.

Chapter 6: Evidence of control caused instability issues that are emerging frequently during the fast-paced inverter-based resource transition to the electricity grid. Grid-forming inverters (GFMIs) are deemed as one of the solutions for small-signal stability issues. This work applies the component connection method (CCM) that cascades unitized state-space models (SSM) into the comprehensive virtual synchronous machine (VSM) model. The model captures a range of frequency response from the switching time delay to power loop control. Dynamics due to VSM's rotating frame disturbances have been included broadly. The time domain simulation shows high accuracy of the assembled model when tracking perturbation responses. The derived frequency model exhibits high similarities to the wideband frequency scanning as well, which opens the potentials for frequency stability analysis in wide-area network.

Chapter 7: Grid-forming control has been appointed as the technology required for achieving a high penetration of renewables in the grid, as it successfully contributes to the power system stability. Doubly fed induction generators (DFIGs) are used widely in wind power generation, so this chapter proposes a grid forming control scheme of DFIGs that achieves its operation as a virtual synchronous machine (VSM). The proposed scheme is based on the rotor flux orientation to a reference axis obtained from the emulation of the synchronous generator swing equation. The rotor flux is oriented to the reference axis by means of a flux controller that also controls the flux magnitude. The flux orientation in turn allows to control the DFIG torque, while the flux magnitude control allows to regulate the generator reactive power or terminal voltage. The proposed control system has been validated using comprehensive simulation models for assessing its grid-forming capability.

Chapter 8: The increasing incorporation of renewable energy in power systems is causing growing concern about system stability. Renewable energy sources are connected to the grid through power electronic converters, reducing system inertia as they displace synchronous generators. New grid-forming converters (GFCs) can emulate the behavior of synchronous generators in terms of inertia provision, power-frequency or voltage-reactive regulation. Nevertheless, as a consequence of synchronous generator emulation, GFCs also present power angle oscillations following a grid disturbance. This chapter proposes two power stabilizers for damping low-frequency oscillations in the power system. The first power stabilizer provides power oscillation damping through the active power (POD-P) using GFCs and the second through the reactive power (POD-Q) using a STATCOM. This work shows the superiority of POD-P stabilizer over POD-Q, but at a cost of employing some kind of energy supply in the DC bus to support the power interchange during system stabilization.

Chapter 9: System stability is investigated for a future Irish grid consisting entirely of GFMIs under three-phase fault conditions with the inverters placed at existing locations for large-scale conventional generation. Electromagnetic transient (EMT) simulations showed that a 100% GFMI system, employing either droop control (or virtual synchronous machine), dispatchable virtual oscillator control, or a mix of both, under a combination of virtual impedance (VI) and scaling current saturation limiting control, is robust against three-phase faults, with consistent performance being achieved, despite variations in fault location or inverter control methods. Freezing GFMIs virtual angular speed during the fault, for both VI and current scaling approaches, system transient stability is greatly enhanced. Time domain simulations also show that when active or reactive current prioritization current saturation controls are applied that GFMIs can introduce large, high-frequency resonance oscillations, but a scaling-down current saturation approach can help to mitigate such problems by generating smoother current references.

Chapter 10: Recently, an emerging communication-free control strategy called dispatchable virtual oscillator control (dVOC) has been introduced for grid-forming inverters (GFMIs). In contrast to droop control that operates based on phasor quantities, dVOC is a non-linear and time-domain approach and stabilizes arbitrary initial conditions to a sinusoidal steady state. As recent studies emphasize the relation between the performance (e.g., stability) of dVOC-based inverters and the Thevenin grid impedance seen by the inverter at its point of common coupling (PCC), it becomes necessary to enable these inverters to online estimate the grid impedance. Therefore, this chapter presents the implementation of an online parametric grid impedance estimation into the control loop of a single-phase dVOC-based inverter operated in a grid-connected mode. The identification of the grid impedance is achieved based on the injection of the pseudorandom binary sequence (PRBS).

Chapter 11: Increasing the penetration of inverter-based generation in a power system results in reduced system inertia, which can lead to various stability issues. As a result, regulation of voltage and frequency are of considerable concern with the increased usage of non-synchronous generation. Grid-forming (GFM) inverters can provide options to help address these challenges. Battery energy storage systems (BESSs) are important for the economic and reliable operation of the grid, because of their capability for energy storage, bidirectional energy exchange, and fast output response. With the increasing penetration of renewable energy sources on the grid, the importance of BESSs is becoming more vital. With an appropriate control strategy, inverters integrated with BESSs can provide a promising solution to stability problems in modern power systems. Topics covered in this chapter include inverter architecture, control architecture, BESS model, integrated BESS with GFM inverter model, large-scale BESS, application example, and simulation results.

Chapter 12: An integrated energy storage system (EMS) that was proposed in this chapter had coordinated control operation of microgrid (MG), which makes the control structure simpler and makes parameter design easier. To provide quick tracking of reference signals, full state-variable model predictive control (MPC) is employed to directly manipulate power switches in the proposed integrated EMS. The suggested scheme's architecture is appropriate for achieving higher performance, straightforward extension and upgrade of DGs in plug-and-play mode, operational reliability, and resilience to communication failure. Using the MATLAB/Simulink® environment, a simulation model of a rural MG with solely renewable energy resource (RER)-based distributed generations (DGs) has been developed. In order to improve system performance in terms of power quality, such as voltage regulation and total harmonic distortion (THD) under steady-state and transient conditions of photovoltaic (PV)-wind generation and sudden load fluctuations, FCS-MPC–based coordinated control has been proposed in integrated EMS for DGs connected via an inverter.

Nabil Mohammed, Hassan Haes Alhelou,
and Behrooz Bahrani

1 Introduction to Grid-Forming Inverters

Akhlaque Ahmad Khan and Ahmad Faiz Minai
Integral University

CONTENTS

1.1 OVERVIEW

The requirement for environmentally friendly power advancements, especially wind and sun oriented, creates inverter-based resources (IBRs) that are turning into an inescapable part of AC power sources. Since these sources are conflicting, IBRs often catch the biggest measure of energy accessible at some random time and feed it back into the framework. IBR inverters are frequently expected to follow network voltages and infuse current into the current voltage. Thus, they're alluded to as "grid-following inverters" (GFLIs). A phase-locked loop (PLL) is a typical approach for synchronizing with the grid voltage. This GFL inverter is behaving like that of a current source. This sort includes practically all right-now introduced IBRs, and thus, voltage source conduct isn't inborn in IBRs [1]. Moreover, IBRs need adequate energy stockpiling to rough inertial reaction. The over-current evaluations of the power electronic exchanging gadgets used in inverters are likewise genuinely low when contrasted with synchronous generators. Thus, IBRs are named non-synchronous generators. Voltage and frequency control will be a key concern as non-synchronous generating sources become more widely used in power systems [2].

Whenever they might work in both grid-connected and islanded modes, microgrids have emerged as a stage for coordinating IBRs [3]. The network controls voltage and recurrence in grid-connected mode, and IBRs only act as GFL inverters. To build a local power grid in islanded mode, one or more inverters should serve as voltage and/or frequency controllers. Answering to this specific need, grid-forming inverters

DOI: 10.1201/9781003302520-1

1

(GFMIs) were created. Furthermore, as the concept of microgrids grew, the need to mimic synchronous generator characteristics arose. As a result, energy storage components and control approaches such as virtual synchronous generator (VSG) operation are being added to GFMIs [4].

Inverter-based sources like photovoltaics (PV) and wind have been accepted to have low idleness since they are normally worked at their appraised power yield and are not supposed to respond progressively to frequency variations [4]. At the point when inverter entrance levels move because of more sustainable power establishments, the absolute put-away mechanical energy falls. Higher frequency motions might create as a greater measure of active energy stockpiling is decommissioned. Bigger variations might cause issues, for example, frequency-based stumbling of burdens and more established framework innovation, which can be dangerous [5]. More modest island power frameworks, like those in Australia and Hawaii, are having idleness issues. This isn't necessary, with increasing penetration, inverters may be adjusted to improve frequency damping. GFMI frequency management has the potential to be highly beneficial, especially for islanded power networks with frequency difficulties. IBRs may modify their output considerably faster than massive synchronous machines, arresting system frequency shifts before any load shedding occurs [6].

Inverter-based technologies were limited to grid-following control strategies in the last decade. In a grid-following engineering, inverters get voltage and frequency estimations from the network along with genuine and reactive power set-focuses from the power station regulator. Thus, the inverter fills in as a modified current source that is adjusted with the grid and answers changes in network voltage and frequency, yet helping the grid's solidarity in any manner can't. In the GFLIs, the power plant controller should always monitor the grid conditions, heavily relying on external set-points from the supervisory system that monitors the power system network. Reliance on a higher supervisory control system introduces communication delays. Therefore, an immediate response that is inherently provided by synchronous generators cannot be mimicked by the GFLI in terms of a fast-reacting generator, not to mention the lack of ability in voltage and frequency regulation in this control approach [7].

Generally, a solar farm power plant is an excellent example of a GFLI application. The solar farm does not respond to grid voltage and frequency changes in a similar way as a coal or steam power plant using a synchronous generator. Solar farm inverters (grid-following) are unable to give inertia to the power grid because they lack the spinning mass to simulate synchronous generator dynamic behavior. Solar farms are frequently linked to a region of the network that is rather weak in strength, necessitating the installation of an extra synchronous condenser at the point of connection to mitigate the effect of inertia. A significant development in battery technologies has made them more cost-competitive with other technologies such as synchronous condensers in recent years. As a result, the importance of battery storage devices in the electrical grid is becoming increasingly widely recognized. This is intensified by a new development in grid-connected inverter control technologies called GFMIs, in which battery energy storage systems are used to establish a grid and maintain its voltage and frequency. A GFMI receives voltage and frequency set-points, which should be maintained on the grid. At the same time, the true and reactive power injection and absorption are proportional to the grid voltage/frequency departure

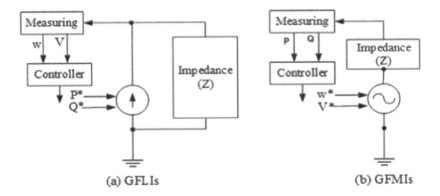

(a) GFLIs   (b) GFMIs

FIGURE 1.1   Block diagram of GFLIs and GFMIs. (Source: [8].)

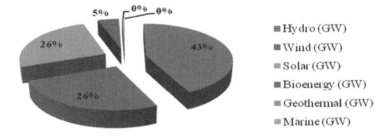

- ■ Hydro (GW)
- ■ Wind (GW)
- ■ Solar (GW)
- ■ Bioenergy (GW)
- ■ Geothermal (GW)
- ■ Marine (GW)

FIGURE 1.2   Worldwide renewable energy generation at the end of 2020 [9].

from specific values. As a result, a GFLI functions as a controllable voltage source. Figure 1.1 portrays block chart of GFLIs and GFMIs. GFM inverters, by and large, answer rapidly and forcefully to voltage and frequency changes in a network, which isn't generally wanted conduct in a powerless network [8].

Global renewable power capacity was estimated by IRENA to be 2,799 GW at the end of 2020. With a capacity of 1,211 GW, hydropower contributed the lion's share of the world total. Figure 1.2 shows how much renewable energy will be produced globally by the year 2020.

1.2   VIRTUAL SYNCHRONOUS GENERATOR

Although GFMIs can form and maintain a grid voltage and frequency, they cannot fully mimic synchronous generator dynamic behavior when changes happen in the network. Any unexpected change in load or the occurrence of a contingency event scan causes the grid voltage and frequency to suddenly deviate from their normal operating levels [10]. In a weak network, this quick change in grid voltage and frequency is a regular occurrence. A generic GFMI has limited ability to increase network robustness due to a lack of inertia.

This concludes that inertia is a crucial factor in any power system network for its resistance to sudden changes and loss of stability. As a result, VSG arose as a novel subset of GFMIs capable of providing synthetic inertia. The term "synthetic" emphasizes that these GFMIs can replicate a synchronous generator's dynamic response without its rotational mass. The VSG responds to the grid frequency change simultaneously, similar to a response caused by an automatic governor control (AGC) action. It reacts to the grid voltage change in the same manner as an automatic voltage regulator (AVR) controller does in a conventional synchronous generator. However, one should consider that the critical fact in implementing VSG inverters is energy source availability. Therefore, battery energy storage should be coupled with the inverter's DC input to provide an uninterrupted energy supply to the VSG. Generally, the battery energy storage state of charge (SOC) for the VSG should remain at 50% SOC to provide adequate headroom for both charging (e.g., during grid frequency rise) and discharge (e.g., during grid frequency drop) [8].

There are significant advantages in using VSG for network strength and stability, especially for a power system network in Australia. In most cases, the VSG functionality is a software add-on rather than a hardware feature. As a result, the cost of converting current battery energy storage systems to synthetic inertia-producing battery energy storage systems may be greatly reduced. VSG will soon be an essential component of grid-connected inverters [8].

1.3   GFL AND GFM INVERTERS

For utility-based inverters, there are two fundamental control schemes: GFLIs and GFMIs. GFLIs infuse a current at a particular stage point to oversee the result of true and reactive power. The grid phase angle is followed continuously utilizing a PLL. The GFLIs are unable to control the system's voltage or frequency directly. A GFMI or the power system provides outside voltage and frequency references. The GFLIs must shut down if the voltage/frequency supply fails. Typical GFL inverters are depicted in Figure 1.3.

GFMIs and GFLIs are not the same things at their core. A GFMI, similar to grid-tied SGs, is a voltage source that is controlled and camouflaged behind a coupling reactance. Direct voltage and frequency control are conceivable with voltage source inverters with droop qualities. During a possibility, grid framing sources with droop the executives will promptly increment or diminish their result ability to adjust stacks and keep up with nearby voltage and frequency. Between increasing output power and changing frequency response, there is no discernible delay. As a result, GFM sources react to any given scenario significantly faster than GFL sources. Using IBRs to provide main frequency control could be very beneficial, especially in power systems with low inertia. Because system frequency fluctuations occur far faster than onboard IBRs, load shedding must be done before system frequency variations arise. In Figure 1.4, a typical GFM inverter is shown.

The O'ahu power system in Hawaii uses a maximum demand summer situation with a total load of around 1.08 GW [12]. In the system, there are 16 synchronous generators with a combined output of 0.66 GW, as well as 0.08 GW of transmission-connected renewables. The rest of the generating fleet, with a total capacity of 0.36 GW, is distributed PV. The departure of a 0.2 GW synchronous generating unit in the framework is the setting off condition. The red "FW" trace in Figure 1.5 depicts the interaction of PV GFLIs with the Frequency-Watt function. The blue "CERTS"

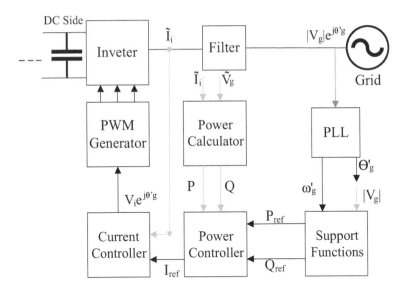

FIGURE 1.3   Block diagram of typical GFL inverter. (Source: [11].)

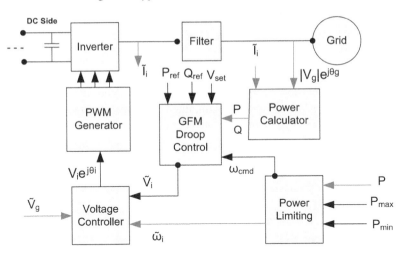

FIGURE 1.4   Block diagram of typical GFM inverter. (Source: [11].)

trace depicts GFMIs' reaction to droop control. The GFM inverters give incredible frequency dampening.

The size of the generation load imbalance suggests that there is sufficient untapped potential or headspace for frequency management [13]. This can be accomplished by the utilization of capacity, turning saves, as well as discontinuous environmentally friendly power sources that are worked at a lower power level than the most extreme achievable power. It is also possible to use batteries that have an inverter interface. One drawback is that idle capacity might lead to greater operational costs [14].

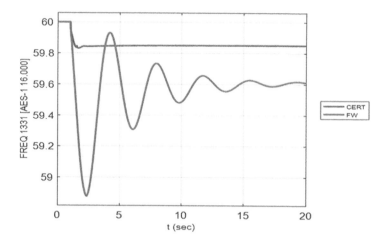

FIGURE 1.5 System frequency regulation with GFLI and GFMI. (Source: [11].)

1.4 COMPARISON BETWEEN GFLIs AND GFMIs

A network bus interfaces the inverter to the remainder of the framework. The inverter on that bus "sees" the grid, and this insight takes the shape of frequency and voltage. The question is, who is responsible for maintaining that voltage and frequency? Whether V/f will still be maintained if the inverter under consideration is disconnected every type of inverter injects power into the bus it's connected to, but the key deciding factor of its role as grid forming, feeding, and supporting is maintaining V/f.

Technically, if the inverter control structure directly incorporates a voltage tracker or regulator and its frequency is auto-generated (not sensed by the bus), then it's a GFMI.

On the contrary, if a voltage tracker is not employed and frequency reference is obtained from sensed/measured/estimated frequency locked to the corresponding bus by a mechanism like PLL or frequency-locked loop, then it's grid following. The evolution between GFL and GFM inverter is shown in Table 1.1, and the performance requirement for grid-forming source is shown in Figure 1.6.

Grid support is an "additional" feature, and a grid-supporting inverter can be of both types: grid forming or feeding. The inverter is said to "support the grid" if the references/set-points of the inverter (set-points include voltage, frequency, true/reactive power, virtual flux, virtual torque, injected currents, and so on) are adjusted based on "additional" inputs from other buses/loads in the network, for serving extra ancillary features (apart from local v/f regulation and energy infusion). Power quality improvement, stability assistance, buffer provisioning, incident ride-through, economic dispatch, and other services are examples of further help [17]. The following are some examples:

TABLE 1.1

Evolution between GFL and GFM Inverters [15–16]

	Grid-Following Inverter	Grid-Forming Inverter
Basic control objectives	Deliver a specified amount of power to an energized grid	Set up grid voltage and frequency
Output quantity controlled	ac current magnitude and phase angle	ac voltage magnitude and frequency
Require a stiff and stable voltage at the terminal?	Yes	No
Control elements present	Compulsorily has a phase-locked loop (PLL)	Compulsorily does not have a phase-locked loop (PLL)
Features	• It controls genuine and receptive power, as well as shortcoming flows. • Incapable of running on its own • Cannot reach a 100% penetration rate	• Balances loads instantly without the need for coordination controls • Self-contained operation • Capable of achieving 100% penetration

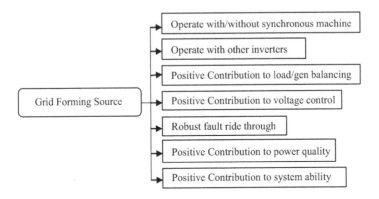

FIGURE 1.6 Performance requirement for grid-forming source. (Source: [16].)

- *Single inverter feeding stand-alone network*: It is solely responsible for generating and maintaining V/f on all buses in the network, resulting in "grid forming". If it adjusts its V/f set-points for other purposes (like brown-out, current limiting for overloads and faults, improving the voltage profile of a remote bus, loss reduction, etc.), then it is also "grid supporting". It is supporting the grid formed by itself! Otherwise, if V/f is regulated at constant values, it is not grid supporting.
- *Parallel-connected droop-controlled inverters*: All inverters contribute or participate in the process of grid formation by means of droop coefficients.

They serve the additional purpose of power sharing; hence, they are both grid forming and grid supporting.

- *PV inverters are always operated at maximum power point (MPP)*: irrespective of the V/f of the bus to which they're connected, they only inject available peak power into the grid, hence only grid feeding [18].
- They inject a certain fraction of available peak power into the grid, and the reference power is adjusted as per the reserve requirement of the grid. Hence, it's an example of both grid feeding and grid supporting.

1.5   FRAMEWORK MODELING

The two-source diminished order power framework model that was utilized in this review to assess the fundamental communications between a generator and an inverter source is displayed in Figure 1.7. A comparative procedure was utilized in a dependability research contrasting the infiltration of GFL versus GFM sources in a microgrid [19].

The inverter source is a brought-together portrayal of all inverter-fed sources in the framework. The model is excessively oversimplified, overlooking highlights such as generator between the region associations and sub-synchronous reverberation. This strategy can be upheld as a first-standards investigation into the principal methods of a blended source framework, which can therefore be extended to incorporate various regions. A model steam generator, comprising of a synchronous machine with a prime mover, an exciter, and a governor, is utilized in this review. The flux decay paradigm with an IEEE Type 1 exciter is utilized to reproduce the SM and AVR regulators. A reheat-type steam turbine is utilized as a gauge for the lead governor, which is addressed as a first-order framework (Figure 1.8). Stator and framework transients are disregarded by the model [20].

The current controller elements of a GFL inverter are generally fast in contrasted with the power reaction and can be overlooked. The power reaction is considered as a first-order framework, and the PLL is addressed as a low-pass filter. Non-linearities in the frequency watt work are overlooked, and a fundamental droop gain is utilized to portray this block (Figure 1.9). The voltage regulator elements of a GFM inverter

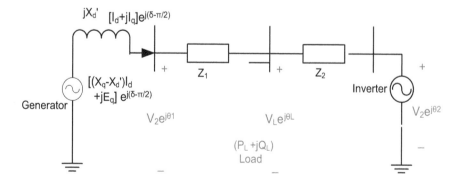

FIGURE 1.7   Single-line diagram of the two-source system paradigm. (Source: [20].)

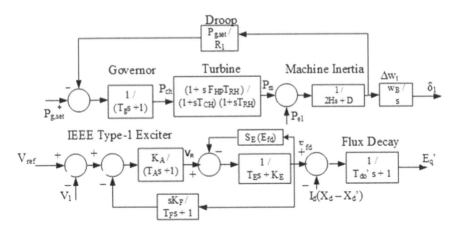

FIGURE 1.8 The generator model's block diagram. (Source: [20].)

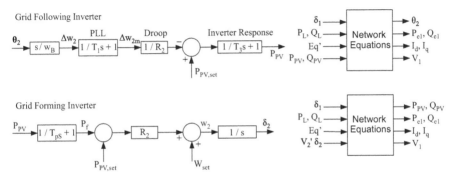

FIGURE 1.9 GFLI and GFMI models. (Source: [20].)

are fairly quick and are disregarded. Just the elements made by the low-pass filter utilized for power estimation are addressed in the GFM inverter. This first system model does not contain the dynamics of the power-limit regulator. The network conditions are given as logarithmic limitations for every circumstance. The prompt infiltration level is the proportion of inverter power supply to add up to load demand. The impedances are scaled by the power result to mirror multiple sources associated in parallel. The grid-forming unit's power output is scaled to allow for easy penetration level change [20].

1.6 METHODS OF GFMI CONTROL

A popular control method for GFLI is vector current control. When utilized as voltage sources, GFMIs, unlike GFLIs, are capable of producing a voltage phasor at their PCC. To guarantee grid synchronization, if fundamental, framework help, the abundancy and angle of the voltage phasor at the PCC are powerfully overseen in the GFMI's internal cascade regulator structure. An internal current control circle, a halfway voltage control circle, a virtual impedance circle, a functioning power regulator (APR), and

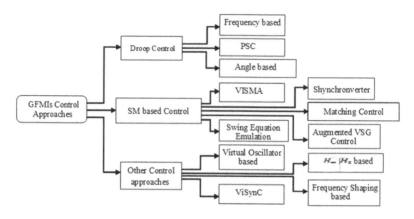

FIGURE 1.10 GFM inverters control approaches. (Source: [1].)

a reactive power regulator (RPR) are among the different inward control circles found in a GFMI. The RPR and APR are utilized to change the amplitude, frequency, and period of the voltage at the PCC. GFMIs can be made without the inward cascade circles by straightforwardly changing the inverter terminal voltage. Kkuni [21] examines the impact of inward cascade circles on GFMI execution. This study takes a gander at frequency and angle control strategies in view of true power, as well as voltage size control techniques in light of reactive power. As illustrated in Figure 1.10, these control approaches may be divided into three types.

1.7 CHALLENGES OF GRID-FORMING INVERTERS

a. *GFMI stability analysis:* In recent years, GFMI stability analysis has garnered a lot of attention. Two sorts of investigations that have been distributed in the writing are small-signal stability analysis and grid frequency stability analysis. More information about the two categories may be found in the subsections below:
   • *Small-signal stability:* Ding [22] offers a technique for investigating small-signal stability in light of areas, which requires an itemized model of the GFMI controller and its boundaries. The area-based technique can be utilized to look into the effects of GFMIs and GFLIs with massive deployment, as well as the consequences of adjusting regulator parameters. However, as the number of GFMIs grows, region-based stability analysis becomes more difficult to accomplish.
   • *Grid frequency stability:* Utilizing electromagnetic transient reproductions, Crivellaro [23] researches grid frequency strength with huge coordination of GFMIs, up to 100%. Greater degrees of penetration, according to reports, could induce interactions with power system stabilizers (PSS), prompting PSS re-tuning to maintain the system stable. Another conclusion is that when GFMIs are widely employed, the grid dynamics drastically alter, influencing the system frequency nadir and the rate of change of frequency (RoCoF).

b. *Switching from stand-alone to grid-connected mode:* Inverters are programmed to inject a specified amount of current into the grid using an MPPT algorithm or a reference given by a central controller in the grid-connected mode of operation. The inverter is operated in a grid-feeding mode, as the reference voltage and frequency are established by the upstream grid. However, in an islanded mode of operation, e.g., a microgrid with local generation and loads, it is crucial to have some of the IBRs operate in a grid-forming strategy to regulate the local voltage and frequency. Hence, a seamless transfer of operation between the GFMI and GFLI modes of operation is essential in such a situation. The two main challenges in obtaining a seamless transfer of operation are fluctuation in frequency and deviation in voltage and current [24].

c. *Fault ride through (FRT) and over-current protection:* GFMIs keep the point of common coupling voltage (vPCC) at its set-point by acting as a voltage source. As a result, during faults and voltage sags, they have to inject a large amount of current into the grid to bring the vPCC back to the set-point. Unlike synchronous machines, which can tolerate up to 6 per unit (pu) of over-current, semiconductor switches inside the GFMI can only tolerate 20%–40% without oversizing [25]. High over-current can lead to the failure of the switches due to the short thermal time constant of the semiconductor [26]. Therefore, a current limitation mechanism is necessary to protect GFMIs from over-currents. As a general rule, these strategies can be arranged into two fundamental gatherings: current-controlled and voltage-controlled limiters.

i. Current limitation:

- *Current-controlled limiter (CCL):* In order to keep the over-current within a tolerable range during a malfunction, current control takes priority over voltage regulation. Grid-following control [2] and current saturator [27] are the two primary subcategories of the current-controlled group. When a failure occurs, the inverter switches to grid-following mode and works as a current source, removing the grid-creating capability. In this mode, the output current of the inverter is controlled to track a predefined fault current waveform [2]. The predefined fault current reference guarantees that the over-current is within a permissible range.

- *The voltage regulator limiter (VCL):* This solution allows the GFMI to operate normally in a voltage-controlled mode. Virtual impedance (VI) is the key element used for limiting over-currents while still operating the GFMI as a voltage source. In the literature, various VI models have been proposed and studied [28].

- *Current limiting for asymmetrical faults:* Depending on the reference frame, the current restriction for asymmetrical faults changes. The natural reference frame (NARF), i.e., the abc-frame, is the most preferable frame for dealing with asymmetrical faults. When an unbalanced fault occurs, the GFMI control is switched to an NARF to independently control each phase current of the inverter [29]. However, this approach requires fault detection and independent control for each phase, thereby increasing the complexity of the GFMI's control.

ii. *GFMI fault recovery:* A variety of parameters, including the control structure and the GFMI's over-current protection, influence the fault recovery
and large-signal transient stability of a GFMI. In terms of the control structure, non-inertial and inertial control can result in different post-contingency
behaviors. In addition, the fault recovery process of a GFMI with either a
current saturator or VI for current limitation is described in [25]. This work
reveals that the VI offers a longer critical clearing time compared to the
current saturator. This result aligns with the analysis in [27]. Besides, the critical clearing time can be extended by limiting the power-angle revolution during a fault. To slow down this revolution, the power references and the droop
gain should be adjusted during a fault as detailed in Taul [30], respectively.

1.8 DISCUSSION

A grid-feeding control is used for a single inverter to deliver power to an already
energized (main) grid at a predefined power (active and reactive) references.

The voltage amplitude and frequency of a solitary inverter in an islanded microgrid
are determined using a grid-forming control. Noted that it does not operate in parallel with any inverter and doesn't share the voltage and frequency regulation responsibility with any inverter.

The basic goal of a grid supporting working as a current source (grid supporting-
grid feeding) is to deliver power (true and reactive). This is a technically advanced
grid-feeding control that makes use of droop control to help with main frequency
and voltage regulation and can work in tandem with other "like-minded" inverters.

A grid-supporting system that functions as a voltage source (grid supporting-grid
forming) does not run in lockstep with a regular grid system. It operates within a
large autonomous microgrid just as a grid-forming controlled inverter. In this case, it
is called grid supporting, not because it is connected to a traditional main grid. It is
called grid supporting because it operates in parallel with "like-minded" inverters with
the same grid-supporting control strategy. With the help of the droop, these inverters
can jointly change the frequency and voltage magnitude of the autonomous microgrid.

1.9 CONCLUSION

As referenced in Taul [30] and Qoria [25], various components of a GFMI's fault
recuperation and enormous signal transient strength are reliant upon various elements of the framework, including the control structure and the GFMI's over-current security. While GFMIs have been successfully implemented at various power
levels in Australia, the United States, and the United Kingdom, their broad usage
requires more technological breakthroughs and regulatory acceptability. The difficulties tended to in this part incorporate small-signal dependability, frequency
regulation strength, over-ebb and flow alleviation, and a consistent progress between
grid associated and free modes. Control strategies for GFMIs are likewise examined. The applied contrasts among GFMIs and GFLIs are talked about, with a focus
on the importance of grid support, particularly in the grid's weaker sections. Areas
that need further investigation and development are also recognized and discussed.
Regulatory support and acceptability, in addition to technological advancements, are
critical facilitators for GFMI adoption in big power networks.

REFERENCES

1. D. B. Rathnayake, M. Akrami, C. Phurailatpam, S. P. Me, S. Hadavi, G. Jayasinghe, S. Zabihi, and B. Bahrani, "Grid forming inverter modeling, control, and applications," *IEEE Power & Energy Society Section (IEEE Access)*, 2021, pp. 114781–114807, doi: 10.1109/ACCESS.2021.3104617.

2. X. Lin, Y. Zheng, Z. Liang, and Y. Kang, "The suppression of voltage overshoot and oscillation during the fast recovery process from load short circuit fault for three-phase stand-alone inverter," *IEEE Trans. Emerg. Sel. Topics Power Electron*, vol. 9, no. 1, pp. 858–871, 2020.

3. H. Han, X. Hou, J. Yang, J. Wu, M. Su, and J. M. Guerrero, "Review of power sharing control strategies for islanding operation of AC microgrids," *IEEE Trans. Smart Grid*, vol. 7, no. 1, pp. 200–215, 2016.

4. Q.C. Zhong and G. C. Konstantopoulos, "Current-limiting droop control of grid-connected inverters," *IEEE Trans. Ind. Electron.*, vol. 64, no. 7, pp. 5963–5973, 2017.

5. A. A. Khan, A. F. Minai, L. Devi, Q. Alam, and R. K. Pachauri, "Energy demand modelling and ANN based forecasting using MATLAB/simulink," *2021 International Conference on Control, Automation, Power and Signal Processing (CAPS)*, 2021, pp. 1–6, doi: 10.1109/CAPS52117.2021.9730746.

6. M. A. Alam, K. Fatima, and A. F. Minai, "Comparative analysis of TPS and EPS of IBDC for power management. In: Iqbal, A., Malik, H., Riyaz, A., Abdellah, K., and Bayhan, S. (eds) *Renewable Power for Sustainable Growth. Lecture Notes in Electrical Engineering*, vol. 723. Springer, Singapore, 2021, doi: 10.1007/978-981-33-4080-0-60.

7. A. F. Minai and A. Tariq, "Analysis of cascaded multilevel inverter," *India International Conference on Power Electronics 2010 (IICPE2010)*, 2011, pp. 1–6, doi: 10.1109/IICPE.2011.5728129.

8. Majid Fard, *Lead Engineer, Power Systems at Aurecon*, 2021. Online available at https://www.linkedin.com/pulse/making-long-story-short-grid-forming followinginverters-majid-fard.

9. International Renewable Energy Agency (IRENA), *Renewable Capacity Highlights 2021*, Newsletter, 31 March 2021.

10. A. F. Minai, T. Usmani, A. Iqbal, and M. A. Mallick, "Artificial bee colony based solar PV system with Z-source multilevel inverter," *2020 International Conference on Advances in Computing, Communication & Materials (ICACCM)*, 2020, pp. 187–193, doi: 10.1109/ICACCM50413.2020.9213060.

11. R. Lasseter, Z. Chen, and D. Pattabiraman, "Grid-forming inverters: A critical asset for the power grid," *IEEE J. Emerg. Sel. Top. Power Electronics*, pp. 2168–6777, 2019, doi: 10.1109/JESTPE.2019.2959271.

12. M. E. Elkhatib, W. Du, and R. H. Lasseter, "Evaluation of inverter-based grid frequency support using frequency watt and grid-forming PV inverter," *Presented at the IEEE Power and Energy Society General Meeting*, Portland, OR, 2018.

13. A. A. Khan, A. F. Minai, Q. Alam, and M. A. Mallick, "Performance analysis of various switching scheme in multilevel inverters using MATLAB/SIMULINK," *Int. J. Curr. Eng. Technol.*, vol. 4, pp. 718–724, 2014.

14. A. F. Minai, T. Usmani, and M. A. Mallick, "Performance analysis of multilevel inverter with SPWM strategy using MATLAB/SIMULINK," *J. Elect. Engg. (JEE)*, vol. 16, no. 4, pp. 428–433, 2016.

15. Y. Lin, J. H. Eto, B. B. Johnson, J. D. Flicker, R. H. Lasseter, H. N. V. Pico, G. S. Seo, B. J. Pierre, and A. Ellis, "Research roadmap on grid-forming inverters," Nat. Renew. Energy Lab., Golden, CO, USA, Tech. Rep. NREL/TP-5D00-73476, 2020.

16. J. Matevosyan, B. Badrzadeh, T. Prevost, E. Quitmann, D. Ramasubramanian, H. Urdal, S. Achilles, J. MacDowell, S. H. Huang, V. Vital, and J. O'Sullivan, "Grid-forming inverters: Are they the key for high renewable penetration?" *IEEE Power Energy Mag.*, vol. 17, no. 6, pp. 89–98, 2019.

17. A. F. Minai, M. A. Husain, M. Naseem, and A. A. Khan. "Electricity demand modeling techniques for hybrid solar PV system" *Int. J. Emerg. Electr. Power Syst.*, vol. 22, no. 5, pp. 607–615, 2021, doi: 10.1515/ijeeps-2021-0085.

18. P. R. Sarkar, A. K. Yadav, A. F. Minai, and R. K. Pachauri, "MPPT Based SPV System Design and Simulation Using Interleaved Boost Converter," *2021 International Conference on Control, Automation, Power and Signal Processing (CAPS)*, 2021, pp. 1–6, doi: 10.1109/CAPS52117.2021.9730712.

19. M. J. Erickson, "Improved power control of inverter sources in mixed-source microgrids," *PhD Thesis,* Univ. Wisc - Madison, 2012.

20. D. Pattabiraman, R. H. Lasseter, and T. M. Jahns, "Comparison of grid following and grid forming control for a high inverter penetration power system," *2018 IEEE Power & Energy Society General Meeting (PESGM)*, 2018, pp. 1–5, doi: 10.1109/PESGM.2018.8586162.

21. K. V. Kkuni, S. Mohan, G. Yang, and W. Xu, *Comparative Assessment of Typical Control Realizations of Grid Forming Converters Based on their Voltage Source Behaviour*, 2021, arXiv: 2106.10048.F [Online]. Available: https://arxiv.org/abs/2106.10048.

22. L. Ding, Y. Men, Y. Du, X. Lu, B. Chen, J. Tan, and Y. Lin, "Region based stability analysis of resilient distribution systems with hybrid grid forming and grid-following inverters," *Proc. IEEE Energy Convers. Congr. Expo. (ECCE)*, 2020, pp. 3733–3740. Virtual, Detroit, United States, Duration: 11–15 October, 2020. doi: 10.1109/ECCE44975.2020.9236196.

23. A. Crivellaro, A. Tayyebi, C. Gavriluta, D. Groÿ, A. Anta, F. Kupzog, and F. Dörfler, "Beyond low-inertia systems: Massive integration of grid forming power converters in transmission grids," *Proc. IEEE Power Energy Soc. General Meeting*, 2019, pp. 1–5. doi: 10.48550/arXiv:1911.02870.

24. M. N. Arafat, S. Palle, Y. Sozer, and I. Husain, "Transition control strategy between standalone and grid-connected operations of voltage source inverters," *IEEE Trans. Ind. Appl.*, vol. 48, no. 5, pp. 1516–1525, 2012.

25. T. Qoria, F. Gruson, F. Colas, X. Kestelyn, and X. Guillaud, "Current limiting algorithms and transient stability analysis of grid-forming VSCs," *Electr. Power Syst. Res.*, vol. 189, 2020, Art. no. 106726.

26. T. C. Green and M. Prodanovic, "Control of inverter-based micro-grids," *Electr. Power Syst. Res.*, vol. 77, no. 9, pp. 1204–1213, 2007.

27 L. Huang, H. Xin, Z. Wang, L. Zhang, K. Wu, and J. Hu, "Transient stability analysis and control design of droop-controlled voltage source converters considering current limitation," *IEEE Trans. Smart Grid*, vol. 10, no. 1, pp. 578–591, 2019.

28. S. F. Zarei, H. Mokhtari, M. A. Ghasemi, and F. Blaabjerg, "Reinforcing fault ride through capability of grid forming voltage source converters using an enhanced voltage control scheme," *IEEE Trans. Power Del.*, vol. 34, no. 5, pp. 1827–1842, 2019.

29. R. Rosso, X. Wang, M. Liserre, X. Lu, and S. Engelken, "Grid-forming converters: An overview of control approaches and future trends," *Proc. IEEE Energy Convers. Congr. Expo. (ECCE)*, 2020, pp. 4292–4299. Virtual, Detroit, United States, Duration: 11–15 October, 2020. doi:10.1109/ECCE44975.2020.9236211.

30. M. G. Taul, X. Wang, P. Davari, and F. Blaabjerg, "Current limiting control with enhanced dynamics of grid-forming converters during fault conditions," *IEEE Trans. Emerg. Sel. Topics Power Electron.*, vol. 8, no. 2, pp. 1062–1073, 2020.

2 Requirements and Grid Standards for Grid-Forming Inverters

Chitaranjan Phurailatpam and Damian Flynn
University College Dublin

CONTENTS

2.1 INTRODUCTION

The chapter focuses on the system-level understanding of grid-forming requirements and the way forward to adopt grid-forming inverters (GFMIs) in power system networks. Firstly, system scarcities and stability issues with high shares of inverter-based resources (IBRs) are discussed, followed by a comparison of the capabilities between grid-forming, grid-following, and synchronous sources to deliver the required system services. A detailed discussion then follows on the performance requirements of various services that can be provided by GFMIs, and the challenges associated with their large-scale implementation. System services capabilities and technical requirements of GFMIs are further classified as mandatory, optional-mandatory, and non-mandatory requirements. Pathways for the standardised adoption of GFMIs in power systems, which can be either through grid code modifications, system services market, or utility (system operator) owned and operated GFMI sources, are then discussed in the following section. With each pathway offering its own advantages and limitations, a combination of these approaches is expected to deliver the best results for specific system requirements. Finally, some of the earliest proposals for grid code modifications and ongoing developments to accommodate GFMIs (Great Britain (GB), Australia, Europe, and the United States) are also covered in detail.

2.2   SYSTEM SCARCITIES AND STABILITY ISSUES WITH HIGH SHARE OF IBRs

With increasing shares of inverter-based renewable energy sources in power systems, conventional synchronous generator-based sources are gradually being displaced by IBRs. Thus, with a reduction in the number of such conventional units being dispatched, the system services that they typically provide, e.g., inertial response, high fault current, synchronising torque, damping power, regulating and contingency reserve, and voltage support, are also being depleted.

Synchronous (or physical) inertia corresponds to the rotating mass of a synchronous generator (hydro/steam turbine) whose speed is magnetically locked with the system frequency. During a frequency disturbance, the stored rotational energy is released/absorbed in the form of an inertial response, thus helping the system to arrest changes in frequency, before slower-acting governor responses, e.g., primary reserve, respond to the frequency deviation. An inertial response is an inherent and instantaneous response that depends on the magnitude of the disturbance, i.e., the total imbalance between generation and demand. Another important classification for inertial response can be termed as *phase jump power*, which is the instantaneous change in power output following a sudden change in the phase angle at the point of interconnection, which may arise, for example, during a system split. This feature is increasingly seen as being important in large, interconnected power systems [1,2].

The ability to (temporarily) supply very high fault currents, which are typically up to 6–8 times their nominal rating, represents an additional important service provided by synchronous machines [3]. Such capability is not only necessary for fault detection and correct operation of power system protection, but a high fault current is also essential for maintaining a voltage profile, which is a fundamental fault ride-through prerequisite [2,4,5]. In addition, synchronous sources provide synchronising torque that helps them to remain in synchronism during rapid changes in voltage angle [5,6]. Finally, damping is provided by the damper winding, following a sudden change in rotor speed, due to a disturbance, helping to reduce local oscillations [1,3,5].

With the gradual displacement of synchronous generators, the system services they have traditionally provided are becoming less abundant, with implications for the behaviour and dynamics of power systems. Historically, numerous standards have been developed for technology specifications and control requirements, linked with synchronous machine capabilities, which have resulted in the existing high levels of reliability and security of supply found in most power systems [1,7–9]. However, the existing fleet of grid-following IBRs, as seen in wind turbines, solar photovoltaic (PV) arrays, battery energy storage systems (BESS), etc., does not cater to most of these essential characteristics and services. Hence, for (present-day/future) scenarios with high shares of IBRs, various stability challenges, including frequency, voltage, rotor angle, and control stability, can be identified [2,3,5,10].

In an interconnected power system, during a contingency, such as a generator tripping, overall system inertia determines the initial rate of change of frequency (RoCoF), while the collective governor response of the generators further helps to arrest the frequency deviation. A reduction in the number of online synchronous generators can lead to higher RoCoFs and higher frequency deviations for the same disturbance size, leading to potential violations in operating limits and triggering of RoCoF and under frequency load shading relays, unless preventative actions are taken [3,11]. In worst-case scenarios, a cascading trip of such protection devices can result in a complete system outage, for example, the 2018 system separation in Australia and the 2019 partial blackout in GB [12,13].

Although some power systems have experienced issues of relatively high RoCoF due to high IBR shares, Ireland and GB, as examples, have an emerging concern of high RoCoF, which can be greater than 0.5 Hz/s due to their relatively small size and limited (synchronous) interconnections [14]. One additional RoCoF-related concern is the loss of embedded generation during faults, which can exacerbate the effects of the original event. Some counteracting measures have been to increase the threshold setting of RoCoF relays, and to phase out vector shift protection, as used for islanding detection [1]. As a longer-term solution, the required system services need to be acquired from multiple and varied sources to maintain stable system operation.

Recognising the possibility of a high share of IBRs located further away from load centres, the maximum power transfer capability across a long distance is determined by the transient angular stability limit, which ensures that machines remain in synchronism during any credible contingencies. Voltage stability, which is associated with the ability of the system to maintain the voltage within a nominal operating range, during both normal and faulted conditions, is also impacted by a high share of IBRs, and the reduction of services that can provide fast injection/absorption of reactive power. Furthermore, with a reduction in system strength, the critical voltage at the nose of the power-voltage curve may fall within the nominal operating range, masking a condition that exceeds the 'reliable' operating point [15].

Control stability [2,5,6,10] is an additional concern, with most IBRs being grid-following in nature. The number of control actions and management layers associated with inverter operation, such as active/reactive power output and terminal voltage, and also controls that are specific to the physical power source, e.g., mechanical torque and wind turbine speed, have become quite complex [2]. Coordination and interoperability of such control layers, both in temporal and spatial distribution, also represent a growing challenge, partially due to (a) a lack of uniform control standards for both GFLI and GFMI, (b) incomplete analytical understanding for developing decoupled control strategies in various timescales with assured system stability, and (c) a lack of transparency of the control hierarchies and mode switching logic embedded within inverter designs from different manufacturers [3,10]. One example of such controller instability was the severe voltage oscillations, with a frequency of $\approx 8\,Hz$ (ranging between 355 and 435 kV on the 400 kV system), that occurred in Scotland during August 2021, when the system was operated with high levels of wind generation and low physical inertia [16]. The problem has been further avoided by dispatching some local conventional synchronous machine-based power plants. Control instability can further be exacerbated by conditions of low system strength, resulting in inaccurate grid voltage measurements and incorrect/rapid current injection from GFLI. Moreover, concerns of sub-synchronous and super-synchronous oscillations have been highlighted with high shares of IBRs, and the need for damping requirements to address the issues [5,6,17].

It follows that in order to overcome the range of system stability issues associated with high inverter (renewables) shares, the required system services must be obtained from a combination of inverter-based and synchronous sources with diverse capabilities. To this end, the capabilities of GFMIs present themselves as being suitable to deliver most of the observed system scarcities, assuming that suitable specifications and requirements are put in place.

2.3 COMPARISON OF GFMI, GRID-FOLLOWING INVERTER, AND SYNCHRONOUS SOURCES: SYSTEM SERVICES PERSPECTIVE

Given the various system scarcities and stability challenges associated with high IBR shares, the prospect of GFMIs addressing the observed depletion of system services has become an attractive solution [5,6,18–24]. The concept of a GFMI is well known, as seen in smaller isolated systems, such as microgrids, and marine applications for quite some time [25,26]; however, applications to large power systems have recently gained popularity due to their characteristics closely resembling those of synchronous machines. The concepts of grid-forming and grid-following are now briefly introduced from the point of view of formulating grid standards, making note of the differences in capabilities between them, and with synchronous sources.

A traditional *grid-following inverter* (GFLI; most IBRs that are currently deployed) requires an external grid voltage measurement to synchronise itself to the grid, remain connected, and adjust its output to track/follow the external grid voltage. The reference waveform and thus the synchronising function of a GFLI is typically provided by a phase-locked loop (PLL) that determines the angle of the grid voltage at the point of connection. In contrast, a GFMI establishes its own internal voltage reference waveform, can synchronise itself with the grid and can adjust its power output to maintain its voltage reference, in a similar manner to that of a synchronous machine. Thus, a GFMI can inherently stabilise grid conditions by providing an instantaneous response to maintain the local voltage and frequency. However, even though GFLIs can provide certain grid support functionality (which can also be sub-categorised as grid-supporting inverters), by adjusting their output, by utilising local measurements of voltage and frequency, the speed of response is limited compared with GFMIs.

A performance comparison of system services that can be provided by GFLI, GFMI, and synchronous machines is presented in Table 2.1. It is important to note that a synchronous condenser (SC), which can either be a newly commissioned bespoke unit or converted from a retired power plant, can also provide most of the listed services provided by a synchronous generator, such as inertial response, high fault current, phase jump power, system strength, synchronising torque, and voltage support. It is also worth noting that the power/current provided by a converter-interfaced source (both GFMI and GFLI), in response to a system disturbance, is typically limited to around 1.2–1.5 times the continuous steady-state rating due to

TABLE 2.1

Performance Comparison of Various Sources for Different Grid Support Services

Service	Grid-Following Source	Grid-Forming Source	Synchronous Machine
System inertia	–	Synthetic – inherent	Physical – inherent
Provide fault current	–	1.2–1.5 pu	6–8 pu
Contribution to phase jump power	–	Yes	Yes
Fast frequency response contribution	Yes	Yes	–
Contribution to system strength	–	Yes	Yes
Provide synchronising torque	–	Yes	Yes
Provide damping power	Limited	Yes	Yes
Blackstart capability	–	Yes	Yes
Contribution to primary frequency response	Yes	Yes	Yes
Voltage/reactive power support	Yes	Yes	Yes

physical limitations of the converter. These limitations apply to all services and capabilities of converter-interfaced sources, creating a clear distinction in capabilities between sources.

In order to further understand performance differences between technologies, the concepts of physical inertia, synthetic inertia, and fast frequency response (FFR) need to be defined and understood. Physical inertia and the corresponding inertial response, as discussed in Section 2.2, are provided by the rotating mass of the synchronous machine. The response is inherent, instantaneous, and does not require any form of measurement or control action. Synthetic inertia can be provided by a converter-interfaced source to mimic the physical inertial response from a synchronous machine. A GFMI can provide instantaneous synthetic inertia during a frequency disturbance without the need for any measurement or additional control implementation. The instantaneous injection, or absorption, of current is due to the voltage angle difference between the internal voltage reference of the GFMI and the network voltage during the disturbance. For a GFLI, a frequency response can be emulated with an additional control layer implemented to respond to the measured system RoCoF, frequency deviation, or a combination of the two [27]. Due to the need for a longer measurement window to reject measurement noise, control delays involved, and activation deadbands associated with the response, there is always a considerable delay (10–100 s of milliseconds [14]) in the response of a GFLI compared to a GFMI. Hence, the frequency response from GFLIs is generally classified as being part of the FFR response.

A FFR, on the other hand, is a control capability that can be implemented in converter-interfaced sources, such as wind turbines, battery storage, high-voltage direct current (HVDC) connection, and deloaded PV systems, to inject or absorb power in response to a measured frequency deviation. FFR will have no, or limited influence, on the maximum RoCoF during a disturbance; however, it will influence the frequency nadir/zenith, and the time to reach the nadir/zenith.

It is also important to note that the reliance on a PLL by GFLIs to remain connected to the grid and to identify and respond to system disturbances, particularly for a low inertia system, is one of the weaknesses of such systems. For example, rapid changes in voltage angle, notably when connected to weaker parts of the grid, can lead to poor tracking performance of the PLL, and loss of synchronism during grid disturbances. Hence, with the current generation of IBRs with no or limited support of the required system services to maintain system stability, the additional benefits provided by GFMIs become quite attractive.

2.4   OVERVIEW OF SYSTEM SERVICES FROM GFMIs AND PERFORMANCE REQUIREMENTS

As part of defining what the term grid forming should mean from a technical performance requirement viewpoint, it is important to assess their (potential) system service capabilities, and how they can contribute to improving system stability and addressing (anticipated) system scarcities. It then becomes possible to identify which technologies can deliver some/all required grid-forming services. Aligning with the services listed in Table 2.1, the associated performance requirements for grid-forming sources to supply them are now considered.

i. *Instantaneous active power, or inertial response, to limit system RoCoF:* Capability to inject/absorb active power instantaneously, based on a change in RoCoF or voltage angle at the point of interconnection. A grid-forming power plant should also possess sufficient energy storage capacity to support a predefined frequency response characteristic, for example, as specified in ECC 6.3.19 [28] and shown in Figure 2.1 (import and export range shown as red and blue envelopes). One important distinction between a GFMI and a GFLI active power response is the ability of the former to react immediately, without any measurement, deadband or control delays. Such a distinction enables a grid-forming source to contribute towards more severe disturbances, such as phase jumps, that can potentially happen during system splits. National Grid ESO's grid code modification GC0137 [1] classifies the overall inertial response power based on the speed of response, as faster 'phase jump power' and comparatively slower 'active inertia power'. The phase jump power is expected to respond within 5 ms and have frequency components exceeding 1,000 Hz. The overall inertial response capabilities of GFMIs are expected to help reduce system RoCoF, and thus the frequency deviation during disturbances.

ii. *Fast fault current injection:* Capability to supply fast reactive current within a very short duration (5 ms in GC0137 [1]) during a fault when the voltage deviates beyond the nominal operating range. As discussed in Section 2.2, such capability is essential for correct operation of protection devices, maintaining the local voltage, and augmenting fault ride-through capabilities of other sources. However, it is important to note that the capacity of

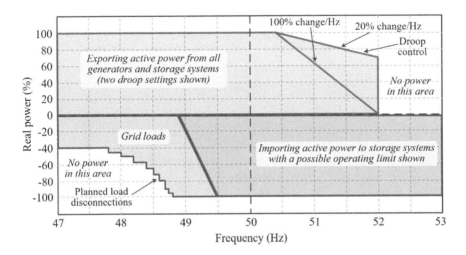

FIGURE 2.1   Power frequency operating characteristics of GFMIs as specified in ECC 6.3.19 [28].

existing converter systems is limited thermally by their electrical components to about 1.2–1.5 times their rated capacity.

Figure 2.2 shows an example voltage-reactive current injection characteristic of a grid-forming source [1]. The GFMI plant (such as generation, HVDC system, and demand) is required to inject no less than its pre-fault reactive current, which must increase if the voltage at the interconnection point falls below 0.9 pu, while ensuring that the overall rating of the GFMI plant is not exceeded. Figure 2.3 shows the injected fault current rating requirement for a GFMI [1], whereby it is required to inject reactive current

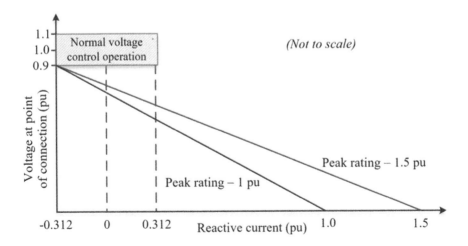

FIGURE 2.2   Voltage-reactive current injection characteristic of grid-forming resource [1].

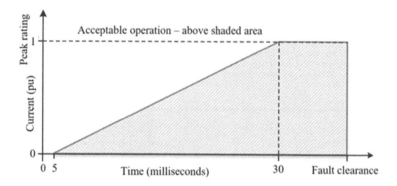

FIGURE 2.3   Injected fault current rating requirement for grid-forming resource [1].

above the shaded area shown in the figure when the retained voltage at the point of interconnection falls to 0 pu.

iii. *Contribution to system strength*: System strength, which also relates to the first two services, can be defined as the ability to generate, maintain, and control the (local) voltage waveform to help maintain stable grid operation, both during steady-state operation and following a disturbance [15]. The capability is important for maintaining stable operation (by providing a voltage waveform reference) in weak grid areas, and potentially offer further headroom for connecting additional grid-following IBRs.

System strength also determines how well a power system can return to normal operation following a disturbance/fault, or how quickly the voltage waveform can be restored. System strength at a given location can be considered to be directly proportional to the fault level at that location (three-phase fault level measured in MVA (million volt-amperes)) and inversely proportional to the effective share of GFLIs [29]. GFMIs can provide fault current as a proxy for system strength and help maintain a stiff system voltage waveform reference to stabilise the output of nearby GFLIs. The Australian Energy Market Operator (AEMO) assesses system strength of a region by setting and reviewing the minimum fault levels needed to maintain power system security across the National Electricity Market (NEM) [29].

iv. *Contribution to damping power*: Capability to damp out power oscillations, especially in the presence of a large share of grid-following sources. Active power damping is expected to be an essential and inherent capability of a GFMI to damp out both sub-synchronous and super-synchronous oscillatory modes.

Sub-synchronous oscillation modes are introduced in power grids due to reactive components of lines/cables (notably in series compensated networks) and electro-mechanical oscillations between synchronous machines (local and inter-area), while super-synchronous oscillations modes can occur due to interactions between the control loops of IBRs [5]. Some of these modes, if poorly damped, can become unstable (negatively damped) when excited by either a small or large disturbance [6]. Hence, damping power, which is the active power that is naturally injected or absorbed to reduce such active power oscillations, is expected to be delivered from grid-forming sources. Quantification of the expected damping requirement in a conventional power system depends on factors such as the number of series compensated lines; however, with a high share of IBRs (both GFMI and GFLIs), the damping requirements are still not well understood and are expected to evolve with time [5].

v. *Contribution to synchronisation torque*: Ability to help maintain synchronous operation between various sources by providing synchronising torque (power) to overcome sudden voltage angle changes during a disturbance.

Rotating machines in a power system are electromechanically coupled to each other, whereby the exchange of power through the network, driven by angular differences between machine rotors, helps to maintain synchronous operation. The power flow that tends to reduce angle differences, and maintain synchronism, is known as the synchronising power, with the corresponding torque, synchronising torque [5]. When the magnitude of this torque reduces with higher shares of IBRs or interconnection across long transmission lines, their ability to exert a positive stabilising influence reduces, decreasing the capability of sources (synchronous machines and IBRs) to ride through and maintain synchronism during transient events, such as grid faults and system splits. Grid-forming sources are expected to provide synchronising torque to help maintain synchronism and improve angular stability with high shares of IBRs; however, the full capability of GFMIs to deliver synchronising torque and the quantification of exact requirements are not fully understood.

vi. *Blackstart capability:* Capability to energise the local network following system blackout conditions and support the system restoration process. GFMIs will be expected to have sufficient capabilities to support/enable a complex blackstart process, which includes initiating the system restoration, supplying balancing power, supporting the local voltage and reactive power needs, and providing sufficient inertia, system strength, primary and secondary frequency response for the duration of the restoration timeline [15]. It is important to note that sources providing blackstart capability not only need to self-start and create a voltage and frequency, but they also need to accommodate a wide range of loading conditions so that load and generation can be added to the island. GFMIs providing blackstart capability need to be designed with sufficient energy buffer and over-current capability to help energise transformers and transmission lines as required. The volume of blackstart service required is not necessarily expected to change, however, depending on grid evolution, rather than a small number of large synchronous machines providing blackstart service, and many small plants (significant portion of which can be GFMIs) are expected to provide such a service in the future [5].

vii. *Support for power system island:* Requirement to sustain island operation with/without synchronous machines (which may occur following a system split) for an extended period. In addition to providing inertia, system strength, voltage support, and primary frequency response, the grid-forming source will also require sufficient energy resources to maintain supply balancing and secondary response. Furthermore, the range of operation (i.e., sufficient control range and ramping capability over real and reactive power output), the stress on the units, the decision-making, and the communications requirement can also be more extreme for GFMIs providing such a service [5]. Portions of a power system that are loosely interconnected and susceptible to islanded operation may also require these services over a longer duration.

The above grid-forming services can be expected to be implemented in various types of sources, such as BESS, wind turbine generators (WTGs), HVDC connections, and deloaded PV systems [30–32]. These sources will generally have some form of energy storage component (or reserve) to support sudden variations in grid conditions. Furthermore, with the addition of energy storage to renewable energy sources without inherent energy storage, to create hybrid power plants, grid-forming capabilities can be potentially implemented and delivered, with the grid-forming capability provided by the renewable source and/or the energy storage.

The technical performance requirements and services capability of GFMIs can also be broadly classified as mandatory, optional, and non-mandatory requirements, as shown in Figure 2.4. Mandatory technical requirements can be considered as the minimal conditions for a source to be categorised and defined as grid forming. Depending on system requirements and the existing (regional) surplus/scarcity of particular system services, capabilities such as blackstart can be categorised as either mandatory or non-mandatory. Furthermore, non-mandatory requirements represent GFMI capabilities that don't essentially distinguish themselves from those provided by a grid-following source. Given the wide variation in size, interconnection capacity, and underlying generation portfolios, etc., system operators in different countries may choose to define quite different minimum performance requirements, with some capabilities being mandatory in one system, but not in another.

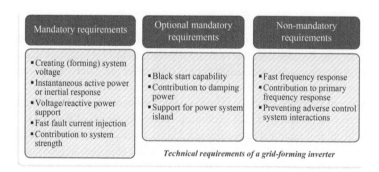

FIGURE 2.4   Mandatory and non-mandatory technical requirements for GFMIs.

2.5   CHALLENGES ASSOCIATED WITH LARGE-SCALE INTEGRATION OF GFMI

Even though the capabilities of GFMIs can be quite attractive, a number of potential challenges face the wide-scale adoption and implementation of GFMIs. One of these is the potential interaction with other (nearby) sources, such as GFLIs, synchronous generators, other grid-forming sources, or the wider system itself. For example, electromagnetic transient (EMT) simulation studies using the IEEE 9-bus test system have shown how interactions between GFMIs and power system stabilisers (PSS) can arise, which might require PSS retuning [33]. Control interactions between grid-forming BESS and offshore wind power plant (grid following) under different grid strengths have also been highlighted [34], which further complicates controller parameter tuning. Consequently, National Grid in GB has suggested using a network frequency perturbation (NFP) plot, combined with a Nichols chart, to assess the possibility of such undue interactions [1]. IEEE draft Standard P2800/D6.3 [35] has also highlighted the use of 'Short Circuit Ratio with Interaction Factors (SCRIF)', which captures the voltage sensitivity between IBRs as a screening tool to identify potential control interaction issues. Furthermore, duality theory has been proposed to analyse similarities and differences between GFMI and GFLI [36], to help understand how instability may arise in networks with high IBR share and necessary measures to combat such instabilities.

The placement and share of GFMIs (compared with GFLIs or other sources) is another crucial factor that can influence stability, especially for large multi-area power systems. The volume/size of GFMIs (with specific services), or the proportion compared with GFLIs, for a particular area of a power system which will sufficiently maintain system stability needs to be carefully evaluated to avoid over-procurement, resulting in costly system operation. Simulation studies of the South-East Australian system have shown that overall frequency stability depends on not only the volume (MW) of GFMI/GFLI resources but also the location of the sources [37]. Another high-level question of concern is whether specifying the GFMI requirements at the point of connection would suffice, or would it be more beneficial to specify internal requirements, such as the source type (or combination) or the range of control topologies.

Many of the stability issues that arise due to high shares of IBR, particularly with a complex mix of GFMIs and GFLIs, cannot be fully analysed and understood with legacy tools/programs, such as phasor-based simulations; hence, there is an increasing need for EMT and state-space–based tools to augment the analysis and design of such systems [6]. Even though the wider frequency bandwidth of an EMT simulation allows for a more accurate and detailed analysis of the faster, more nonlinear, and unbalanced system conditions, full-blown EMT simulations of large power systems can be highly computationally expensive and time-consuming. Moreover, the issue of performing such stability analyses is further aggravated by the hidden nature of the control systems, and mode switching of converters designed by different manufacturers, resulting in the possibility of system operators requiring/mandating detailed EMT models to be submitted for identifying any unstable operating conditions.

The lack of grid-scale testing of grid-forming capabilities represents a broader challenge. Most real-world test systems involve smaller systems (for example, the 720 kVA/500 kWh grid connected BESS at EPFL [38] and the 2.3 MVA grid-forming test setup in NREL [6]), with testing limited to a few specific capabilities, not to mention the lack of testing capabilities with high shares of IBRs (both GFMIs and GFLIs). In this regard, the full capabilities of GFMIs cannot be fully assessed and understood without thorough grid-scale testing [15]. Furthermore, waiting until very high shares of IBRs (predominantly GFLIs) are reached to identify and understand the problems, and then imposing retrospective GFMI requirements later, rather than developing the GFMI capabilities to ensure future reliability, may not be the best approach, as the cost of inaction can be rather steep and prohibitive [6]. This is a limitation that many power system operators throughout the world have recognised, and a number of collaborative projects between industry, researchers, and system operators are underway to carry out necessary testing.

Once the capabilities of GFMIs have been fully understood, system operators can decide upon the minimum requirements for a source to be deemed as grid forming. In order for a source to qualify and provide grid-forming services, certain testing specifications, simulation models, and monitoring data requirements will need to be designed and developed to enable wide-scale adoption of GFMIs.

2.6 PATHWAYS FOR WIDE-SCALE ADOPTION OF GFMI IN POWER SYSTEM

Assuming that the range of services to be provided by GFMIs has been defined, and the large-scale integration challenges have been clearly identified, a range of different implementation pathways can still be followed, as shown in Figure 2.5, each with their own advantages and disadvantages. The options range from (a) grid code modifications (the 'stick' approach), whereby minimum technical requirements are specified for grid-forming sources, to (b) system services markets (the 'carrot' approach), whereby financial incentives are introduced to encourage (gird-forming) inverters to provide certain capabilities, and finally (c) utility-owned and operated

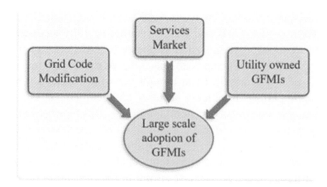

FIGURE 2.5 Pathways for wide-scale adoption of GFMIs in power system.

GFMIs, whereby the utility takes responsibility for ensuring certain critical system capabilities. A particular pathway is not considered 'better' than another, and different system operators may choose different approaches based on historical practice, existing plant portfolio capabilities, etc. Indeed, a combination of approaches may be adopted, whereby some services are implemented in the form of grid code modifications, and others in the form of new service markets, or alternatively, basic capabilities might be imposed by a grid code, with enhanced performance incentivised through service markets. It should also be noted that grid code modifications, new system services arrangements, etc., should be technology-agnostic, wherever possible and appropriate, and hence open to grid-following and synchronous machine-based alternatives.

a. *Grid code modifications:*

Grid connection codes, in general, are technical requirements, regulations, and behaviour that apply to all active participants in the power system, including power generators, adjustable loads, storage, and other units, and they play a critical role in ensuring security of supply. Grid codes typically evolve and adapt over time to define requirements based on plant size, interconnectivity, expansion plans, existing capabilities, IBR share, etc. For systems seeking to achieve a high (variable) renewables share, the role of GFMIs needs to be emphasised through grid codes [39], particularly if 100% renewable systems are envisioned.

This pathway aims to define a certain set of standards and requirements for a source to be classified as GFMI. Grid code modifications will seek to implement a minimum technical specification within the grid code for parties providing a grid-forming capability, with particular categories of plant (perhaps based on plant size (MW) or grid connection voltage (kV)) mandated to provide certain capabilities. Minimum/mandatory capabilities, as discussed in Section 2.4, for a specific power system can be tailor-made and should be defined in detail.

Figure 2.6 shows how grid codes can help to deliver innovation trends, such as the adoption of GFMIs, which will eventually lead to adoption of higher renewable shares [39]. In addition to the adoption of new standards, grid codes can also ensure harmonisation and interoperability between regional grid codes (including distinct transmission system and distribution system grid codes), enable the formation of system service markets (but without defining the system services themselves, e.g., by requiring frequency and voltage control capabilities, and/or requiring that certain plant signals are recorded and communicated), and even tackling cyber-security concerns. A typical high-level grid code requirement is shown in Figure 2.7, which is defined according to specific system needs. While basic requirement specifications are often similar, significant variations in chosen parameters and the range of grid user facilities can apply, based on country and site specifications. Identifying appropriate parameters for suitable user facility classes according to system needs is a crucial part of grid code design. The parameter settings may actually be given in various appendices

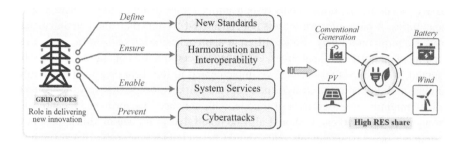

FIGURE 2.6   Role of grid codes in delivering new innovation trends [39].

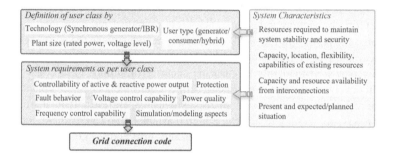

FIGURE 2.7   High-level grid-code requirement specifications according to system needs [39].

such that they can be later revised without the need to update the grid code text itself.

b. *System services markets:*

Grid codes can often require that generation technologies possess certain capabilities, e.g., frequency/voltage control, often as part of future-proofing against anticipated future system needs, but not necessarily that those capabilities are actually provided to the power system. In some systems, IBRs can participate in providing system services, such as FFR, and provide reactive power and frequency regulation support, in competition with synchronous-based, and other, alternatives. Services that are specific to GFMIs can also be defined, and the detailed definition of such capabilities can essentially lead to the creation of a market structure. The creation of services markets, in turn, is then expected to drive the required innovations to enable wide-scale adoption and integration of GFMIs.

It is important to note that system services, such as inertia, fault level, synchronising torque, etc., have historically been provided 'free of charge' by synchronous generation units, given that they represent inherent properties of such plant which cannot be withdrawn, and were mostly available in excess, and hence not highly valued. Certain capabilities, such as regional blackstart support, may have been remunerated separately for eligible plant.

However, with the displacement of such synchronous sources, and given that these capabilities are not inherent to IBRs, such services can be incentivised to be offered in order to maintain a minimum level of security and stability of the system. It is also noted that grid code-based approaches tend to result in minimum performance requirements being met, but not exceeded. In contrast, remuneration-based approaches can promote innovation, particularly if additional weightings are applied based on the degree to which a minimum performance requirement is exceeded.

c. *Utility or system operator owned GFMIs:*

Unlike the above approaches, where GFMIs can (only) be owned and operated by private entities, an alternative pathway suggests that grid-forming functionality is sufficiently critical to secure and reliable power system operation that it cannot be reliant upon the investment decisions of external actors. Consequently, the utility/system operator should be responsible for suitably ensuring that the present day and future system has sufficient (locational) grid-forming capability, provided, in part, or perhaps in full, by GFMIs. In this case, the required upfront cost for research and innovation is taken up by the system operator, with the implementation cost ultimately suitably shared across all electricity consumers.

Such an approach can have some specific advantages in certain circumstances. For example, the size and placement of grid-forming sources can play an important role in the overall system stability. In such cases, utility-owned and operated sources can be selectively placed in the grid based on existing (and anticipated future) stability concerns. Furthermore, the quality of grid-forming services can be uniformly standardised, with the available capabilities and restrictions fully understood, as they will likely not be designed and manufactured by a disparate range of private entities, further reducing the effort to understand the capabilities of multiple different GFMIs and the possibilities for local and system-wide interactions.

With the overall objective of ensuring safe and secure power system operation, and at minimum cost to the end user, the above pathways offer their own advantages and disadvantages, in terms of driving innovation, reducing risk, and ensuring minimum capabilities. For most systems, it is likely that a balance will be chosen between the grid code and markets approach, with potentially different decisions made for different required capabilities, while the utility-owned approach will be a backup option, particularly noting the potential interference with electricity market operations, given that many grid-forming technologies can also participate in energy, system service, and capacity markets.

2.7 STATE OF PLAY AND ADVANCEMENTS IN GFMI GRID CODES

Before investigating the role of grid-forming requirements within existing, and future, grid codes, it is imperative to acknowledge how GFLIs are represented within present day requirements. For example, although the existing generation of

FFR services from IBRs (including frequency deviation and RoCoF-based response from BESS) could, in theory, be integrated at scale with the required performance to improve system stability, policies, grid codes, and economic factors still present barriers for wider application and investment [14]. To this end, EirGrid/SONI in Ireland, as one example, has developed a multi-year program ('Delivering a Secure, Sustainable Electricity System' – DS3 [40]), which defines two new fast frequency services, namely, FFR and synchronous inertial response (SIR). FFR operates in the time period of 2–10 s (although faster responses are incentivised) and is open to all qualifying synchronous and non-synchronous sources, while SIR represents the inertial response from synchronous sources only, in the time frame of 0–5 s. In a similar manner, PJM (Pennsylvania-New Jersey-Maryland interconnection) in the USA defines a frequency regulation service, 'Dynamic Regulation Signal (RegD),' which is specifically designed for fast ramping IBRs such as BESS [41]. Therefore, it is crucial to note that the grid-supporting potentials of GFLIs are fully exploited before placing (perhaps onerous) requirements to make converters grid forming.

Even though grid code regulations do commonly exist for wind turbines, PV solar, BESS, and other inverter-based technologies in many systems, the grid-following (or indeed grid-forming) nature of the converters is rarely explicitly acknowledged. Instead, the focus is placed on PCC performance in terms of voltage, fault ride-through capability, and frequency/voltage control. It is possible that the development of grid code regulations for GFMIs will essentially supersede existing requirements made with respect to GFLIs, which will become a subset within future grid code regulations designed for GFMIs. The following sub-section discusses ongoing developments in GB, Europe, Australia, and the United States as part of developing and establishing (consolidated) grid code requirements for the mass adoption of GFMIs.

a. *National grid ESO grid code modification GC0137 – Great Britain*

National Grid ESO has acknowledged that the natural capabilities provided by conventional synchronous generation in contributing to system stability will no longer be available in sufficient volume, and, in the future, they will have to be paid for [1]. The aim of grid code modification GC0137 [1] is to complement National Grid's stability pathfinder programme [42], which looks for the most cost-effective way to address stability issues and to incorporate a minimum non-mandatory grid-forming specification within its grid code. Subsequently, this will form the foundation for a future short-term stability market which would sit alongside other balancing services, such as the fast-acting Dynamic Containment (DC) frequency response service [43] (rapid response delivered outside ±0.2 Hz, to contain frequency within ±0.5 Hz). The modification seeks to implement a technology-agnostic minimum specification for parties wishing to offer grid-forming capability, given that the plant provides similar services that are traditionally associated with synchronous generators.

Participation in, and the benefits of, the grid code modification will be open to all technologies, such as converter-based generation, 'smart' loads, storage systems, and other novel technologies, which could even include wide-scale vehicle-to-grid (V2G) schemes, or even traditional synchronous generator (or condenser) plants, which already satisfy the requirements of the proposed specification. Three main topics are addressed within the modification, namely, technical performance requirements, as summarised in Table 2.2, plant data and modelling information, and compliance information to demonstrate grid-forming capabilities.

Even though the capabilities are labelled differently by different system operators, most basic definitions remain the same, as discussed in Section 2.4. 'Active control-based power' refers to a plant's ability to change its active power output by adjusting the available power set-points, with a bandwidth below 5 Hz to avoid resonance. GC0137 recommends a phase jump limit of 5° (phase angle between internal voltage of the GFMI and grid connection point), in which a GFMI is still able to inject active phase jump power without reaching the maximum converter limits, and a 60° phase jump angle withstand capability, where the GFMI should ride through the phase jump event without disconnecting. The GFMI should also have the capability to provide active inertia power at the point of connection to the grid, with sufficient storage capacity to meet the frequency response characteristics as specified in ECC 6.3.19 (Figure 2.1). Both active phase jump power and inertial response power should respond naturally within less than 5 ms, with the former having a faster frequency response than the latter. National grid ESO also specifies that a GFMI should provide active damping power with a damping factor between 0.2 and 5. Furthermore, fault current injection from a GFMI should be delivered within 5 ms when the voltage falls below 90% of its nominal value (as shown in Figures 2.2 and 2.3). The grid-forming active power specified in GC0137 includes the active inertia power, active phase jump power, and active damping power.

The next main specification of the GC0137 grid-code modification is the data requirement, which will ensure that the developer provides the appropriate data and models for the GFMI plant, so that the system operator can perform software stability analysis and enable market participation. Data requirements include the grid-forming plant ratings and the key performance parameters, as listed in Table 2.2, while the modelling requirement includes a high-level architecture of the plant and an equivalent simulation block diagram model. The final specification provides compliance requirements in three main areas, which include simulation, testing, and online

---

TABLE 2.2

Key Technical Performance Requirements of a Grid-Forming Plant as Specified in National Grid's Grid Code Modification – GC0137

i) Active control-based power	ii) Active phase jump power
iii) Active damping power	iv) Active inertial power
v) Active RoCoF response power	vi) Control-based reactive power
vii) Fast fault current injection	viii) Grid-forming active power
ix) Grid-forming capability	x) Voltage jump reactive power

monitoring, which is designed to ensure that the plant is capable of meeting the full requirements of the grid code.

b. *National electricity market (NEM) – Australia*

AEMO prepared a white paper [15] on the application of advanced grid-scale inverters in the NEM in August 2021. Advanced grid-scale inverters are defined as IBRs that can provide capabilities such as system strength, disturbance withstand capability, inertial response, primary frequency response, support for a power system island, and initiate/support system restoration, with the majority of these requirements not relevant for a GFLI. The white paper represents the initial step in exploring advanced inverter technologies, such as GFMIs, by adopting a capability and application-led approach to describe the functionality required from advanced inverters. Four different applications were identified as being relevant, in order of increasing capability (relating to complexity/difficulty in being achieved), which are expected to grow in relevance with technology maturity and evolving system needs, are listed as follows:

i.   *Connecting IBR in weak grids:* Maintain stable operation in areas with high shares of IBRs and potentially provide system strength to support connection of other nearby IBR plant.

ii. *Supporting system security:* Provide capabilities such as inertia and system strength, which were predominantly provided by synchronous machines.

iii. *Islanding operation:* Capability to maintain stability and supply balancing to support areas of the grid that have temporarily become separated from the main synchronous network.

iv. *System restart:* Capability to energise and support the local network following a system blackout and assist with the restoration process.

In the same white paper, AEMO recommended prioritising the deployment of grid-forming capabilities on grid-scale BESS, as this technology provides the capability to deliver firm, flexible energy behind the inverter. Furthermore, batteries (with storage capacity of several minutes), coupled with a variable renewable energy (VRE) plant, might also provide a flexible resource mix to cater to the same applications.

c. *ENTSO-E's perspective – Europe*

The European Network of Transmission System Operators for Electricity (ENTSO-E) has also investigated grid-forming requirements for the European power system and has defined seven capabilities that could be termed grid forming if fulfilled in their entirety [44].

i.   Creating (forming) system voltage.

ii. Contribution to fault level (short circuit power).

iii. Contribution to inertial response (limited by energy storage capacity and the available power rating of the Power Park Modules (PPM) or HVDC converter station).

iv. Support system survival to enable effective operation of low-frequency demand disconnection following rare system splits.

v.   Acting as a sink to counter harmonics and inter-harmonics in system voltage.

vi.   Acting as a sink to counter any unbalance in system voltage.

vii.   Prevent adverse control system interactions

In order to facilitate system-level integration of grid-forming sources, ENTSO-E suggests that the required capabilities be defined in connection network codes or CNCs (category of European grid codes that defines the connection requirements of generators/loads/HVDC connection) to enable harmonised solutions, preferably through mandatory non-exhaustive requirements, to begin and accelerate the process of grid-forming implementation. HVDC, FACTS, and SCs are immediate candidates to explore the development of grid-forming capabilities. ENTSOe also suggests that storage and sector coupling facilities should be further explored in order to unlock their full potential for providing grid-forming capabilities.

d. *United States perspective*

The Universal Interoperability for Grid-Forming Inverters (UNIFI) Consortium, which is supported by the US Department of Energy (DOE) and headed by National Renewable Energy Laboratory (NREL), was formed in 2021 to overcome various issues and challenges for wide-scale adoption of GFMIs [45]. The main objective of UNIFI is to understand how a 100% inverter-based system at the scale of the Western Interconnection (~276 GW in capacity in 2019) can be achieved. The consortium is tasked with creating standard specifications for inverter manufacturers and guiding grid operators on its control and operation.

In November 2020, NREL published a 'research roadmap' for GFMIs [4], with a focus on examining inverter controls, assessing their impact on grid stability, and evaluating crucial system interactions (protection systems, electro-mechanical, and IBRs). NREL identified the lack of an established body of experience for operating a hybrid power system, with a significant amount of GFMI and GFLI resources on the scale of the North American interconnection. The research roadmap concludes by offering a multi-layer perspective towards wide-scale GFMI adoption. The North American Electric Reliability Corporation (NERC), in December 2021, also announced considerations for bulk power systems with grid-forming technology, in the form of a white paper [46]. GFMI and GFLI capabilities were compared, along with their major performance characteristics and advantages. The white paper also provided recommendations for North American entities to study as part of deploying GFM technology, while supporting system reliability and resilience despite increasing IBR shares.

Finally, Energy Systems Integration Group (ESIG), which is a non-profit educational organisation, has suggested that, rather than being locked in a circular chicken-and-egg problem between defining requirements for grid-forming capability and the development of that capability, an evolving system needs perspective is presented [6], which is summarised in the outer loop of Figure 2.8 as steps 1 through 9. Firstly, the target system is defined in terms of energy quantities, expected shares of different sources, and expected sinks. The desired resilience against certain disturbances is subsequently determined, before the type and scope of the minimum system needs are determined in step 3, followed by technical performance requirements that are defined for necessary system services. Next, the system services are quantified, followed by identification of economically optimal service provision in steps 5 and 6. In step 7, technical benchmarking is developed and specified to verify performance at the commissioning stage, while the final two steps specify implementation and performance monitoring. The three inner steps in Figure 2.8 show how the nine (outer) steps relate to IBR equipment manufacturers, project developers, and owners.

Figure 2.9 summarises an overall comparison of current advances in GFMI standards and grid codes in various power systems or jurisdictions across the world. A comparison of GFMI capabilities/services has also been introduced to understand the subtle differences between different counties. The pathways being adopted or planned for implementation of GFMIs, which include the grid-code modification

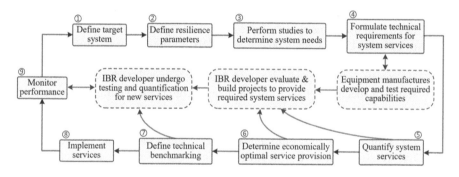

FIGURE 2.8   ESIG's proposed process for deploying new GFMIs [6].

Power system or jurisdiction	GFMI services defined	Grid code modification	Services market	Initiatives & projects	Status & prospects
National Grid ESO (GB)	A, B, C, D, E & G	☑	■	☑	Non-mandatory grid code modification proposed
ENTSO-E (Europe)	A, B, C, D, F & H	■	●	☑	Implementation to depend on location/urgency
AEMO (Australia)	A, B, C, D, F & H	■	■	☑	Grid-forming capabilities in grid scale BESS by 2025
United States	■	■	●	☑	Establish standards and begin adoption in ~ 5 years

A – Inertia and phase jump power
B – Fault current injection
C – Contribution to system strength
D – Voltage/reactive power support

E – Contribution to damping power
F – Power system island
G – Blackstart capability
H– Reduce harmonics & control interactions

☑ Available / adopted
■ Planned / upcoming
● Not planned / unknown

FIGURE 2.9   Comparison of current developments made for wide-scale adoption of GFMIs.

and ancillary services market routes. Finally, the existing status and prospects for each country are summarised. It can be observed that different system operators are at different stages along the GFMI development and implementation process, with noticeable differences in the offered services and approach, given that GFMIs are still at an early stage of adoption. However, with ever-maturing GFMI technology and a deeper understanding of system needs, the differences between standards and the developed grid codes that are currently seen may well largely disappear.

2.8 CONCLUSIONS

GFMIs present themselves as a viable solution for integrating high shares of renewable energy sources into power system networks while maintaining existing levels of system security and stability. Existing challenges with high share of IBRs have been highlighted, while recognising the capabilities of GFMIs to supply depleted system services, followed by discussions on pathways for its adoption, and some advances in the development of grid codes for GFMIs. It is crucial to note that the capabilities of GFLIs have not yet been fully exploited, before enforcing additional GFMI requirements. Considering that some island and isolated systems are progressing rapidly with the deployment of the current generation IBRs, the time to develop and implement the advanced capabilities is quickly running out. Moreover, investors/investments in developing GFMI capability demands certainty, as retrofitting requirements at a later stage will always be more expensive than incorporating features in the design phase. Hence, as recognised and acknowledged by many system operators, continuous development and collaboration work is required to identify the full capabilities of GFMIs and to develop standards or grid code requirements towards their wide-scale adoption in future power systems.

REFERENCES

[1] "GC0137: Minimum specification required for provision of GB grid forming (GBGF) capability (formerly virtual synchronous machine/VSM capability)," National Grid ESO, Nov 2021 2021. [Online]. Available: https://www.nationalgrideso.com/.

[2] J. Matevosyan et al., "A future with inverter-based resources: Finding strength from traditional weakness," *IEEE Power and Energy Magazine,* vol. 19, no. 6, pp. 18–28, 2021, doi: 10.1109/MPE.2021.3104075.

[3] L. Meegahapola, P. Mancarella, D. Flynn, and R. Moreno, "Power system stability in the transition to a low carbon grid: A techno-economic perspective on challenges and opportunities," *Wiley Interdisciplinary Reviews: Energy and Environment,* vol. 10, no. 5, p. e399, 2021.

[4] Y. Lin et al., *Research Roadmap on Grid-Forming Inverters,* National Renewable Energy Lab. (NREL), Golden, CO (United States), 2020.

[5] J. Bialek et al., *System Needs and Services for Systems with High IBR Penetration,"* Global Power Systems Transformation Consortium (G-PST), 2021. [Online]. Available: https://globalpst.org/system-needs-and-services-for-systems-with-high-inverter-based-resource-penetration/

[6] J. Matevosyan and J. MacDowell, *Grid-Forming Technology in Energy Systems Integration,* ESIG - Energy Systems Integration Group, March 2022. [Online]. Available: https://www.esig.energy/wp-content/uploads/2022/03/ESIG-GFM-report-2022.pdf.

[7] G. Strbac, D. Kirschen, and R. Moreno, "Reliability standards for the operation and planning of future electricity networks," *Foundations and Trends® in Electric Energy Systems,* vol. 1, no. 3, pp. 143–219, 2016.

[8] NERC Reliability Standards, North American Electric Reliability Corporation, 2022 https://www.nerc.com/pa/Stand/Pages/default.aspx (accessed 2022).

[9] National Electricity Transmission System Security and Quality of Supply Standard, National Grid ESO, 2019. [Online]. Available: https://www.nationalgrideso.com/document/141056/download.

[10] Q. Peng, Q. Jiang, Y. Yang, T. Liu, H. Wang, and F. Blaabjerg, "On the stability of power electronics-dominated systems: Challenges and potential solutions," *IEEE Transactions on Industry Applications,* vol. 55, no. 6, pp. 7657–7670, 2019.

[11] B. Hartmann, I. Vokony, and I. Táczi, "Effects of decreasing synchronous inertia on power system dynamics—Overview of recent experiences and marketisation of services," *International Transactions on Electrical Energy Systems,* vol. 29, no. 12, p. e12128, 2019.

[12] AEMO, "Final report-Queensland and South Australia system separation on 25 August 2018," AEMO Information & Support Hub, Tech. Rep., Australia, 2019.

[13] Technical report on the events of 9 August 2019, National Grid ESO, Warwick, UK, 2019.

[14] L. Meng et al., "Fast frequency response from energy storage systems—a review of grid standards, projects and technical issues," *IEEE Transactions on Smart Grid,* vol. 11, no. 2, pp. 1566–1581, 2019.

[15] Application of advanced grid-scale inverters in the NEM," Australian Energy Market Operator (AEMO), White Paper, 2021.

[16] D. B. Watson, *How Renewables Caused Scottish Grid's Double Heart Attack*, 2022. [Online]. Available: https://eandt.theiet.org/content/articles/2022/02/how-renewables-caused-scottish-grid-double-heart-attack/.

[17] D. Flynn et al., "Technical impacts of high penetration levels of wind power on power system stability," Advances in Energy Systems: The Large-scale Renewable Energy Integration Challenge, pp. 47–65. Hoboken, NJ, USA: John Wiley & Sons, 2019.

[18] H. Zhang, W. Xiang, W. Lin, and J. Wen, "Grid forming converters in renewable energy sources dominated power grid: Control strategy, stability, application, and challenges," *Journal of Modern Power Systems and Clean Energy,* vol. 9, no. 6, pp. 1239–1256, 2021.

[19] M. Ndreko, S. Rüberg, and W. Winter, "Grid forming control scheme for power systems with up to 100% power electronic interfaced generation: A case study on Great Britain test system," *IET Renewable Power Generation,* vol. 14, no. 8, pp. 1268–1281, 2020.

[20] R. H. Lasseter, Z. Chen, and D. Pattabiraman, "Grid-forming inverters: A critical asset for the power grid," *IEEE Journal of Emerging and Selected Topics in Power Electronics,* vol. 8, no. 2, pp. 925–935, 2020, doi: 10.1109/JESTPE.2019.2959271.

[21] A. Crivellaro et al., "Beyond low-inertia systems: Massive integration of grid-forming power converters in transmission grids," in *2020 IEEE Power & Energy Society General Meeting (PESGM),* 2020, pp. 1–5, doi: 10.1109/PESGM41954.2020.9282031.

[22] H. Holttinen et al., "Design and operation of energy systems with large amounts of variable generation: Final summary report, IEA Wind TCP Task 25," 951388757X, 2021.

[23] H. Holttinen et al., "System impact studies for near 100% renewable energy systems dominated by inverter based variable generation," *IEEE Transactions on Power Systems,* vol. 37, no. 4, pp. 3249–3258, 2020.

[24] P. Christensen et al., "High penetration of power electronic interfaced power sources and the potential contribution of grid forming converters," *European Network of Transmission System Operators (ENTSO-E),* 2020. [Online]. Available: https://

eepublicdownloads.entsoe.eu/clean-documents/Publications/SOC/High_Penetration_
of_Power_Electronic_Interfaced_Power_Sources_and_the_Potential_Contribution_
of_Grid_Forming_Converters.pdf

[25] J. Rocabert, A. Luna, F. Blaabjerg, Rodri, x, and P. guez, "Control of power converters in AC microgrids," *IEEE Transactions on Power Electronics,* vol. 27, no. 11,
pp. 4734–4749, 2012, doi: 10.1109/TPEL.2012.2199334.

[26] E. Hossain, E. Kabalci, R. Bayindir, and R. Perez, "Microgrid testbeds around the
world: State of art," *Energy Conversion and Management,* vol. 86, pp. 132–153, 2014,
doi: 10.1016/j.enconman.2014.05.012.

[27] H. Karbouj, Z. H. Rather, D. Flynn, and H. W. Qazi, "Non-synchronous fast frequency reserves in renewable energy integrated power systems: A critical review,"
International Journal of Electrical Power & Energy Systems, vol. 106, pp. 488–501,
2019, doi: 10.1016/j.ijepes.2018.09.046.

[28] European Connection Conditions (ECC), National Grid ESO, 2022. [Online]. Available:
https://www.nationalgrideso.com/document/114841/download.

[29] System strength in the NEM explained, Australian Energy Market Operator (AEMO),
2020. [Online]. Available: https://aemo.com.au/-/media/files/electricity/nem/system-
strength-explained.pdf.

[30] V. Gevorgian, S. Shah, W. Yan, and G. Henderson, "Grid-forming wind: Getting ready
for prime time, with or without inverters," *IEEE Electrification Magazine,* vol. 10, no. 1,
pp. 52–64, 2022, doi: 10.1109/MELE.2021.3139246.

[31] B. Pawar, E. I. Batzelis, S. Chakrabarti, and B. C. Pal, "Grid-forming control for solar
PV systems with power reserves," *IEEE Transactions on Sustainable Energy,* vol. 12,
no. 4, pp. 1947–1959, 2021, doi: 10.1109/TSTE.2021.3074066.

[32] E. Rokrok, T. Qoria, A. Bruyere, B. Francois, and X. Guillaud, "Classification and
dynamic assessment of droop-based grid-forming control schemes: Application in
HVDC systems," *Electric Power Systems Research,* vol. 189, p. 106765, 2020, doi:
10.1016/j.epsr.2020.106765.

[33] A. Tayyebi, D. Groß, A. Anta, F. Kupzog, and F. Dörfler, "Interactions of grid-forming power converters and synchronous machines," *ArXiv Preprint,* arXiv:1902.10750,
2019. [Online]. Available: https://arxiv.org/abs/1902.10750[34] F. Zhao et al., "Control
interaction modeling and analysis of grid-forming battery energy storage system for
offshore wind power plant," *IEEE Transactions on Power Systems,* vol. 37, no. 1,
pp. 497–507, 2021.

[35] IEEE Draft Standard for Interconnection and Interoperability of Inverter-Based
Resources (IBR) Interconnecting with Associated Transmission Electric Power
Systems," P2800/D6.3, December 2021, pp. 1–181, 2021.

[36] Y. Li, Y. Gu, and T. C. Green, "Rethinking grid-forming and grid-following inverters:
A duality theory," *arXiv preprint arXiv:2105.13094,* 2021.

[37] B. K. Poolla, D. Groß, and F. Dörfler, "Placement and implementation of grid-forming
and grid-following virtual inertia and fast frequency response," *IEEE Transactions on
Power Systems,* vol. 34, no. 4, pp. 3035–3046, 2019, doi: 10.1109/TPWRS.2019.2892290.

[38] Z. Yuan, A. Zecchino, R. Cherkaoui, and M. Paolone, "Real-time control of battery
energy storage systems to provide ancillary services considering voltage-dependent
capability of DC-AC converters," *IEEE Transactions on Smart Grid,* vol. 12, no. 5, pp.
4164–4175, 2021, doi: 10.1109/TSG.2021.3077696.

[39] Grid codes for renewable powered systems - IRENA, IRENA, 2022. [Online]. Available:
https://www.irena.org/publications/2022/Apr/Grid-codes-for-renewable-powered-
systems.

[40] DS3 system services protocol - regulated arrangements DS3 system services implementation project, SONI and EirGrid, 2019.

[41] P. Interconnection, "Implementation and rationale for PJM's conditional neutrality regulation signals," Tech. Rep., 2017 [Online]. Available: https://www.pjm.com/~/media/committees-groups/task-forces/rmistf/postings/regulation-market-whitepaper.ashx

[42] Network Option Assessment (NOA) Stability Pathfinder, *National Grid ESO.* [Online]. Available: https://www.nationalgrideso.com/future-energy/projects/pathfinders/stability.

[43] "Dynamic Containment (DC), " *National Grid ESO.* [Online]. Available: https://www.nationalgrideso.com/balancing-services/frequency-response-services/dynamic-containment?overview.

[44] Grid-Forming Capabilities: Towards System Level Integration, *European Network of Transmission System Operators for Electricity (ENTSO-E)*, 2021. [Online]. Available: https://vision2030.entsoe.eu/.

[45] Will Grid-Forming Inverters Transform Renewables?, *IEEE Innovation at Work.* [Online]. Available: https://innovationatwork.ieee.org/.

[46] Grid Forming Technology - Bulk Power System Reliability Considerations (White paper), NERC, 2021. [Online]. Available: https://www.nerc.com/comm/RSTC_Reliability_Guidelines/White_Paper_Grid_Forming_Technology.pdf.

3 Power System Requirements for Grid-Forming Converters

Mohamed Younis and Hoda Youssef
Independent Electricity System Operator

CONTENTS

3.1 INTRODUCTION: BACKGROUND AND DRIVING FORCES

Prior to connecting any facility to a public grid, the connected facility must meet the minimum technical requirements of this grid. These requirements are essential to reliably and securely operate the electric grid and plan for the future. Grid connection requirements may change from one jurisdiction to another or even within a single jurisdiction. In some cases, there are multiple layers of requirements for the same area, (a) local system requirements, (b) regional system requirements, e.g., Northeast Power Coordinating Council (NPCC) and Western Electricity Coordinating Council (WECC), and (c) continental system requirements, e.g., North American Electric Reliability Corporation (NERC), and European Network of Transmission System Operators for Electricity (ENTSO-E).

Traditionally, the requirements are split into two separate sets of requirements [1] based on the point of interconnection (POI) with the utility, either high-voltage transmission or low-voltage distribution systems. Both sets are developed and updated by the regulating authority that is responsible for protecting the system integrity and

DOI: 10.1201/9781003302520-3

network operation such as Public Utility Commissioners (PUCs) and/or system operators. Due to the significant impacts of distribution systems on their upper stream bulk grid, Regional Transmission Organizations (RTOs) and Independent System Operators (ISOs) have included more requirements to allow for more coordination and interoperability between transmission and distribution systems [2]. Many renewable resources have been connected to the grid in both transmission and distribution systems and more resources are planned to be connected. The majority of the new renewable resources are connected to the power system grid through power electronic converters. Multiple grid challenges may arise due to the introduction of these resources, especially with high penetration levels compared to conventional synchronous resources. However, these resources can be used to alleviate these issues by enabling their capabilities, e.g., grid forming.

Many utilities, especially those with high penetration levels of power electronic resources, have developed their grid code requirements to reliably integrate more wind and solar resources. However, these grid code requirements, with the exception of National Grid ESO [3], do not include any provision for the grid-forming capability of the power electronic resources, commonly known as grid-forming Inverter-Based Resources (IBRs).

This chapter provides an overview of the main performance requirements for connecting IBRs to the public grid, focusing on grid-forming technology. The requirements may significantly change from one system to another due to multiple factors such as system size, historical reliability level requirements, and penetration levels of IBRs. The requirements are grouped based on their impacts and relationship to each other. For example, the Frequency Response and Control section includes inertia, primary frequency and droop control, fast frequency response (FFR), and needs for energy reserve/storage to enable grid-forming capability subsections. The main system requirements include: (a) frequency response and control, (b) active power requirements, (c) fast fault current injection and short circuit contribution, (d) fault ride through requirements, (e) reactive power requirements and voltage regulation, and (f) black start and island operation.

3.2   FREQUENCY RESPONSE AND CONTROL

The system frequency response depends on the system inertia response, primary frequency control, fast frequency control, and secondary frequency control of the machines and resources connected to the system. This section provides a summary of these features and their minimum requirements.

3.2.1   INERTIA

Inertia is defined as the tendency of an object to resist changes in its velocity or motion. In conventional power systems, inertia can be defined as the stored kinetic energy in the rotating masses that are directly connected to the power system. This stored kinetic energy resists any changes in the rotating speed or system frequency. Rotating masses are the directly connected machines' rotors, turbines, and rotating exciters in the power systems. The value of the inertia of a system may significantly change, even during the same day, as it depends on the online rotating masses, i.e., connected and synchronized to the grid at the time of the frequency event. Operating a power system with low inertia could cause frequency excursion during a large generation load imbalance, e.g., loss of a large generation, and it makes the power system more prone to transient instability. Therefore, inertia is considered one of the key reliability services that help maintain the system's frequency and thus its stability. The inertia constant of a machine, H, is defined as the kinetic energy divided by the MVA (million volt-amperes) rating of this machine [4]:

$$H = \frac{\text{Kinetic Energy}}{MVA_{rating}} = \frac{\frac{1}{2}J\omega_{om}^2}{MVA_{rating}} \left[second\right] \tag{3.1}$$

$$\frac{\partial f}{\partial t} = \frac{\Delta P * f_o}{2H} \left[\frac{Hz}{second}\right] \tag{3.2}$$

where
    J: moment of inertia of the entire rotating mass in kg.m^2,
    ω_{om}: rated mechanical speed of the rotation in radian/second,
    $\partial f / \partial t$: rate of the change of the frequency (RoCoF) in Hz/second,
    ΔP: per unit of MW generation load imbalance, and
    f_o: nominal frequency, i.e., 50 or 60 Hz.

The typical range of the inertia constant, H, for a conventional synchronous machine is 2–10 seconds [4].

As discussed above, the natural system inertia depends on the inertia of synchronized rotating machines connected to the grid. For large interconnected grids, the inertia is expected to be high and it is unlikely to experience any issues. However, smaller systems shall pay more attention to their inertia level. The North American grid provides an example of the different system sizes [5], and it consists of multiple independent interconnections that are connected through high-voltage direct current (HVDC) systems as shown in Figure 3.1. The HVDC systems inherently do not provide or allow a transfer of inertia. The largest interconnection is the Eastern Interconnection, which covers central Canada to the east coast and the east of the Rocky Mountains to northern Texas and Florida. This massive system has very large inertia compared to Texas or the Electric Reliability Council of Texas (ERCOT) and Quebec. ERCOT has calculated the critical system inertia below which the system may experience significant frequency excursion and instability [6]. They developed a real-time inertia monitoring tool to continuously calculate the online inertia and compare it to the minimum threshold value. The operators require more synchronous generations to come in service if the system inertia falls below the critical value. These stringent inertia requirements are not only specified in ERCOT but also in other relatively small interconnections with inertia concerns such as National Grid ESO. Currently, the large Eastern Interconnection does not need such requirements but some individual ISOs start to develop inertia monitoring tools to prepare for the future.

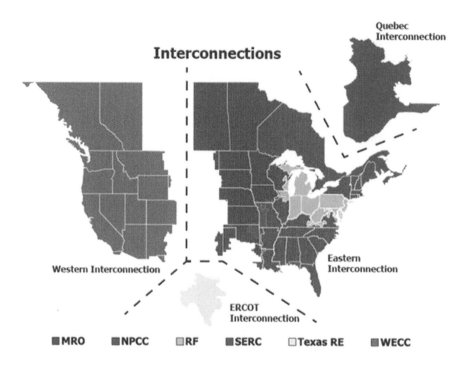

FIGURE 3.1 NERC interconnections and reliability regions [5]. (Source: This information from the North American Electric Reliability Corporation's website is the property of NERC and available on NERC's website in the About NERC Key Players section. This content may not be reproduced in whole or any part without the prior express written permission of the North American Electric Reliability Corporation.)

Many utilities experience high levels of integration of converter-interfaced resources and loads due to different reasons such as the significant reduction in the cost of power electronic converters and renewable resources, decarbonization initiatives, and policies. The converters inherently do not provide any inertia support to the system and the resources behind the converter are completely decoupled from the rest of the grid. Therefore, the inertia has started to decline and the system reliability may deteriorate [7]. Figure 3.2 shows the frequency response and the corresponding powers of each stage. The impact of low system inertia on RoCoF and frequency nadir is shown in Figure 3.3.

3.2.2 PRIMARY FREQUENCY AND DROOP CONTROL

The resources start to respond to the generation load imbalance and frequency deviation following the inertial response as shown in Figure 3.2. Traditionally, this autonomous response due to the droop control is given by conventional synchronous generators with turbine-governor control. However, IBRs are required to provide this response, especially for power systems with high penetration levels of IBRS. The resources shall be able to regulate their output active power with an average droop which can be adjusted between a certain range, e.g., 3%–7%.

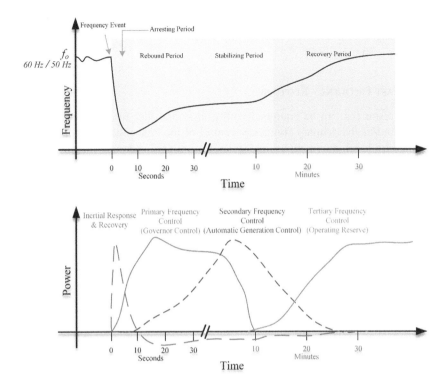

FIGURE 3.2   Frequency response after an event and the corresponding powers [7].

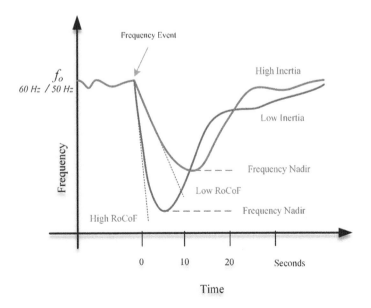

FIGURE 3.3   RoCoF and frequency nadir under different system inertia [10].

The maximum deadband of the droop control shall be reasonable to allow a fair share among all generations during any frequency event, e.g., not wider than ±0.06% or ±36 mHz [8,9].

3.2.3 Fast Frequency Response

The IBR resources can be equipped with a fast controller to emulate the inertial response and/or the primary frequency control of the conventional resources. This FFR shall be local, autonomous, and quick (less than 2 seconds) [11] to provide a similar response (or better) to the rotating machines equipped with a droop control. A fast RoCoF can be used as a trigger event to activate this feature. The resources need to have energy reserve or headroom to be able to provide a sustained response and this point is discussed further in the next section.

3.2.4 Needs for Energy Reserve/Storage to Enable Grid-Forming Capability

For grid-connected resources, it is essential to have energy storage to provide sustained frequency support to the grid. For conventional rotating machines that are directly connected to the grid, the stored kinetic energy can instantaneously provide support to arrest the frequency drops. However, IBRs do not inherently have this feature due to the decoupling characteristic as explained earlier. The IBRs must have a means of energy storage to provide momentarily and primary frequency support. Most renewable IBR facilities, wind, and solar resources are operated using a Maximum Power Point Tracking (MPPT) control method to maximize their power production [12]. Although this method ensures the economic operation of their facility, it does not leave any room for supporting the system frequency, e.g., injecting active power, when the frequency drops. For wind turbines, the inverters may use the stored kinetic energy in their blades to support the declining frequency if this feature is available and activated. This power boost is limited, and it requires a recovery period, i.e., absorbing power from the grid after it finishes. Some ISOs [9] set requirements for this temporary boost in active power such as (a) at what frequency this power boost shall be triggered and canceled, (b) how long shall it takes to be triggered and last, (c) how long it takes to be ready again for the next disturbance, (d) minimum power to be injected, and (e) the rate of change of energy for withdrawing and injecting to the grid [9]. Figure 3.4 shows the power boost of a wind facility and the system frequency for two different cases [9,13]. If the wind speed is low pre-event and this feature is activated, the wind blades may slow down below the cut-out speed. In this case, the wind generation would shut down and exacerbate the situation instead of supporting the system frequency [14].

Solar photovoltaic (PV) resources do not have rotating masses, and therefore they are not able to provide power boosts similar to those of wind resources. To solve this drawback, an energy buffer can be used by curtailing a portion of output power and injecting it into the system when needed, i.e., during frequency decline events. However, this solution may limit the value of renewable resources and reduce their utilization factor. The hybrid generations can be used to provide energy reserve by combining multiple generations such as intermittent renewable generations, wind or solar, with a Battery Energy Storage System (BESS). This combination can not only provide frequency support but also result in a dispatchable generation with a smoother output power.

> Resources that are connected to the grid shall provide the following minimum frequency response and regulation:
>
> i. An inertial response or equivalent response to the conventional synchronous machine with similar power ratings, and
> ii. A droop control adjusted within the applicable range and deadband.
>
> These responses shall be autonomous and local.

3.3   ACTIVE POWER REQUIREMENTS

As discussed in the droop requirements, the resource shall be able to automatically, without any intervention of operators, change their active power to respond to the frequency changes. In case the frequency event continues for a few minutes, the system operators may require a resource to manually change their active power to alleviate a system issue, e.g., thermal overload. The resources shall comply with the active power requirements within their capabilities. For instance, resources, conventional generations or IBRs, at their rated injected active power or at their maximum available injected active power are not expected to increase their active power further, i.e., no headroom. However, all resources shall be able to reduce their active power as quickly as possible or within the applicable time frame. Resources at their minimum active power, or batteries at their maximum charging state, are not expected to reduce their active power or absorb any more active power from the grid, respectively.

> Resources shall be able to continuously control their active power between the minimum and maximum output power.
>
> The facilities shall follow any instructions to change their active power from the system operator within their capabilities.

3.4   FAST FAULT CURRENT INJECTION AND SHORT CIRCUIT CONTRIBUTION

During faults in power systems with conventional synchronous generations, the generator current can rise to six times the rated current during the sub-transient period [15]. Then, the generator current gradually decays to the steady-state fault current which is about 2–4 times the rated current. The synchronous generator has an external DC source that supplies the field current. The existence of the field source and the prime mover, that keeps rotating the generator shaft, helps to maintain the generator's internal induced voltage in the stator windings during the fault. After the terminal voltage drops, the generator's Automatic Voltage Regulator (AVR) tries to compensate for this drop by increasing the voltage of the excitation system. The maximum voltage that the excitation system can reach is called the exciter voltage ceiling. The reactive power of the synchronous generator is expected to increase during the fault. In contrast, the electrical active power of the synchronous generator is expected to decrease and may reach zero if the fault is located at the generator terminal voltage. The simplified equation for the electrical active power of the synchronous generator is [4]:

$$P_e = \frac{EV_t}{x}\sin\delta \qquad (3.3)$$

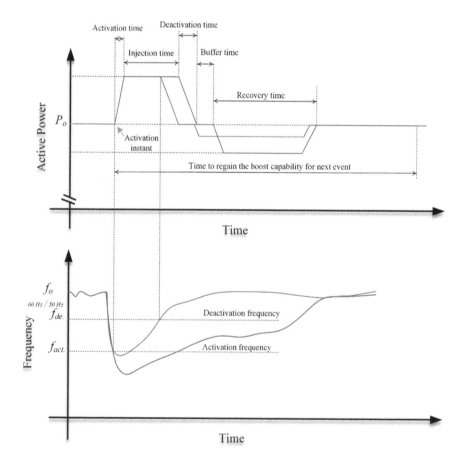

FIGURE 3.4 Power boost of turbine blades and system frequency response [9,13].

where

P_e: synchronous generator active power,

E: steady-state, transient or sub-transient generator internal voltage,

V_t: generator terminal voltage,

x: synchronous, transient or sub-transient reactance, and

δ: load angle in radian.

In opposite to the synchronous generator behavior, the IBRs provide a minimal fault current contribution [15], usually settling in a range of 1.1–1.2 of the rated current of the inverter after an instantaneous spike, and this fault current is fully controlled. To support the grid voltage and protection, the inverters shall inject reactive current at their maximum short-term capability and as fast as possible, e.g., one to three cycles [16]. This fast reactive current injection should be able to help recover the system voltage during the fault, and it should be enough to trigger the system protection. After clearing the fault, the inverters shall block this current and return to the pre-fault settings. The maximum short-term current of the inverters depends on the inverter components and mainly the semiconductor switches. Using inverters with higher short-term current capability would allow a high fault current for longer periods, therefore more support to the voltage and easier fault detection for traditional protection systems. However, this feature adds more cost to the inverter hardware [17]. The high fault current would also enhance the grid strength and this is in contrast to the traditional inverters that weaken the grid due to their minimal contributions to the short circuit current.

Resources with grid-forming capability shall be able to prioritize the reactive injection during the fault. This feature is enabled when the voltage at the POI drops below a certain threshold, e.g., 0.8 pu. The converters shall be able to inject a current with a peak equal to multiple times the rated current. The current shall reach its peak as fast as possible, and it is allowed to decay to its rated current after a certain period or the fault is cleared and the voltage is recovered.

3.5 FAULT RIDE-THROUGH

During disturbances, the system usually faces voltage and/or frequency excursions. The resources, conventional generations and IBRs, are required to stay connected to the system as long as these voltage and frequency excursions are within acceptable ranges. Improper tripping of the resources may lead to adverse impacts on system stability due to cascading failures. Fault ride-through is divided into voltage and frequency requirements for both extreme cases, high and low. The ride-through characteristic has three main zones: (a) normal operation zone, (b) must remain connected or no trip zone, and (c) may trip/disconnect zone. The normal operation zone specifies the range in which the resource can continuously operate for an indefinite time. In the must remain connected or no trip zone, the resource shall not trip within the specified time durations. The resource may remain online to support grid reliability to the best extent possible based on equipment limitations in the may trip/disconnect zone. Some manufacturers included another operating mode for inverters, other than trip and stay connected modes, known as momentary cessation. Momentary cessation means that the inverters block the firing commands of the power electronic switches. Therefore, there is no active current and reactive current. The inverter can re-inject current once the voltage recovers to a certain value, e.g., 0.5 pu. It is recommended to eliminate the momentary cessation operating mode to let the IBRs help the grid during the disturbance. However, if this mode cannot be eliminated, it is recommended to reduce their settings to the lowest values possible [12].

3.5.1   Voltage Ride-Through

All resources, conventional generations and IBRs, are required to remain connected to the grid during any event that includes a voltage deviation. The resources cannot be designed to withstand all grid events without limitations. Therefore, minimum voltage ride-through (VRT) requirements shall be determined based on the system and respected contingencies. However, these requirements shall be reviewed as the penetration level of IBRs increases and equipment capability improves. The VRT is required for both transmission- and distribution-connected resources. Figure 3.5 shows an example of VRT requirements for transmission-connected resources, i.e., NERC PRC-024 Generator Frequency and Voltage Protective Relay Settings [18].

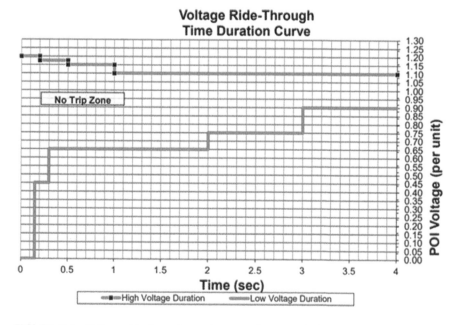

FIGURE 3.5   Voltage ride-through requirements for transmission-connected resources in North America [18]. (Source: This information from the North American Electric Reliability Corporation's website is the property of NERC and available on NERC's website in the United States Mandatory Standards Subject to Enforcement section. This content may not be reproduced in whole or any part without the prior express written permission of the North American Electric Reliability Corporation.)

IEEE Std 1547 includes VRT requirements for distribution-connected IBRs or Distributed Energy Resources (DERs). This standard was developed in 2003 as a voluntary guideline for RTO/ISO; however, after two years, it became mandatory in the United States under the Energy Policy Act [19]. The standard has evolved through the years from requiring the DERs to trip during system disturbance to allowing the DERs to ride through and recently requiring the DERs to ride through in the 2018 amendment [20]. Currently, the IEE 1547 standard has three levels of VRT requirements based on the penetration level of DERs [20].

3.5.2 FREQUENCY RIDE-THROUGH

The frequency ride-through (FRT) requirements are similar to those of VRT. All resources, conventional generations and IBRs, are required to ride through frequency excursions during abnormal system events. During an under-frequency event, the system issue is exacerbated by a loss of IBRs due to improper under-frequency ride-through requirements. The NERC PRC024-2 has a FRT requirements diagram as shown in Figure 3.6.

> Resources shall have the capability to withstand voltage and frequency excursions that happen during system disturbance. This capability must, at least, meet the ride-through requirements.

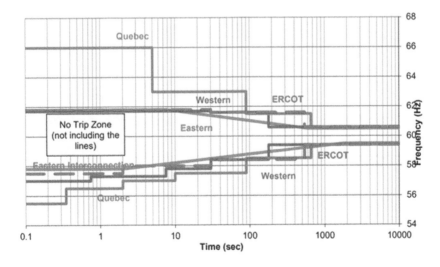

FIGURE 3.6 Frequency ride-through requirements for transmission-connected resources in North America [18]. (Source: This information from the North American Electric Reliability Corporation's website is the property of NERC and available on NERC's website in the United States Mandatory Standards Subject to Enforcement section. This content may not be reproduced in whole or any part without the prior express written permission of the North American Electric Reliability Corporation.)

3.6 REACTIVE POWER REQUIREMENTS AND VOLTAGE REGULATION

In contrast to active power requirements, reactive power requirements can be different from area to area. The reactive power requirements are specified at the POI of the resource to the transmission system. This point is usually the high-voltage side of their Main Output Transformer (MOT), but in some cases, the POI can be in the upper stream transmission system. For some systems, the reactive power requirements are variable and depend on the resource's active power. For example, the reactive power requirements are 1/3 of the active power, i.e., 0.95 lag/lead Power Factor (PF) at POI [9]. Other systems use a fixed reactive power requirement under all active power conditions [21]. Figure 3.7 shows the reactive power capability curve for a resource/battery.

One important aspect of the reactive power requirement is the response time. Normally, the reactive power requirements are continuous and dynamic, i.e., fast response. Some jurisdictions specify a range for static and dynamic reactive power requirements [21]. If the resource facilities are required to increase their static reactive power, a properly sized capacitor bank shall be used. If the resource facilities are required to increase their dynamic reactive power, a properly sized Flexible AC Transmission Systems (FACTs) device, e.g., Static VAR compensator (SVC) or A STATic synchronous COMpensator (STATCOM), shall be used [22].

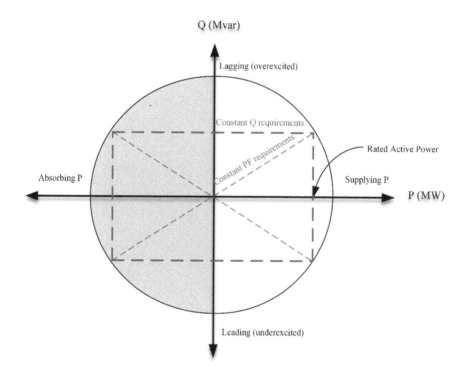

FIGURE 3.7 Reactive power capability curve and requirements.

Resources shall be able to control the voltage at the POI within a certain range of the nominal voltage, e.g., ±5% or 95%–105%. Some jurisdictions allow a range of accuracy resolution for voltage control, e.g., ±0.5%, and an acceptable range of response time or equivalent time [9].

The generation facility shall be able to provide the reactive power requirements at the POI. The reactive power requirements can be provided in different forms: minimum reactive power value, minimum PF or P-Q curve for both directions, i.e., leading and lagging.

3.7   BLACK START AND ISLAND OPERATION

Selected resource facilities are used to restore and re-energize the grid after an event of a partial or a complete shutdown of the system, i.e., a blackout or a brownout. This feature is not required to be implemented in all resources connected to the grid, and it requires a bilateral agreement/contract between the resource facility and the ISO.

The black start facilities must comply with the black start grid requirements to be able to reliably provide the black start service. To understand the black start requirements, we need to start with what is happening during the restoration procedures. After a widespread blackout or brownout and the loss of the electric supply, the ISO calls for the black start facilities to re-energize the grid. The facilities must be able to start up on their own, i.e., without any outside electrical supply and within the agreed time frame. After starting up, the facilities shall start to energize the transmission lines to other generation units, known as cranking path, and they must have the capability to withstand the sustained and switching transient overvoltage due to line charging currents of the unloaded lines. The high voltage is expected to be higher than the normal voltages, around 105% of the nominal voltage, but not higher than the emergency limit, around 110% of the nominal voltage [23]. Then, the facilities shall start to gradually pick up loads in steps, and the system is expected to experience a drop in the frequency and power oscillations. Similar to the voltage, the black start facilities should be able to ride through this abnormal frequency excursion and power oscillations. After starting up more generation units and creating an island(s), the ISO starts to synchronize the islands to form a larger and more stable system. After energizing their areas, the ISOs extend their help to energize their neighbors and regions. These main procedures for system restoration are documented in the system restoration plan. The system operators use the system restoration plan during the restoration of a large system outage, and the plan is regularly updated and reviewed. The data and information in the restoration plan are considered confidential, and they are only shared with the restoration participants, including the black start facilities.

From the procedures above, the black start facilities must comply with the black start requirements, which include: (a) the active power capability to supply the pickup loads and other generation auxiliary loads and withstand the instantaneous loading, (b) the reactive power capability to supply the large charging current, (c) fast voltage and frequency response/control to stabilize them within acceptable ranges,

(d) adequate voltage and FRT characteristics, (e) power oscillation damping, (f) self-starting with applicable time frame, (g) following the restoration plan procedures and instructions including regular tests, exercises and drills, and (h) readiness for subsequent starting up after restoration failures and tripping, i.e., system re-collapse. For microgrids and small systems, similar requirements can be considered but on a much smaller scale and complexity.

In the past, the conventional synchronous generations, with different fuel types, were the only black start facilities for large power system restoration. However, grid-forming resources have started to be implemented, in limited numbers, as black start facilities in some areas. Hornsdale Power Reserve is an example of a lithium-ion battery storage system that provides a black start service to the South Australia system [24]. Moreover, the limitations in grid-forming resources for supplying high charging currents can be mitigated by using a soft start method during the starting up. This method, also known as soft energization, starts up the resource at low voltage and gradually ramps up the voltage of the grid to avoid high charging currents, transient overvoltage, and restoration failures [25].

Prior to participating in the black start service, the facility shall provide evidence through regular field tests, offline assessments, and restoration drills that

 i. the facility can start without any assistance from the grid and within the applicable time frame,
 ii. the facility is able to withstand and alleviate large voltage or frequency excursions that might occur during the system restoration including load pickup and line energization.

The facility shall have adequate power to restore the designed zone at any time. The source of this power can be a resource or stored energy.

REFERENCES

[1] Sutherland, Peter E. "Canadian grid codes and wind farm interconnections." *2015 IEEE/IAS 51st Industrial & Commercial Power Systems Technical Conference (I&CPS).* Calgary, AL: IEEE, 2015.
[2] Electricity Advisory Committee (EAC), "The transmission-distribution interface recommendations for the U.S. department of energy." 2018.
[3] National Grid ESO, "GC0137: Minimum specification required for provision of GB grid forming (GBGF) capability (formerly virtual synchronous machine/VSM Capability)." 2021.
[4] Kundur, P. *Power System Stability and Control,* New York: McGraw-Hill, 1994.
[5] North American Electric Reliability Corporation, "Interconnection." 2022. [Online]. Available: https://www.nerc.com/AboutNERC/keyplayers/PublishingImages/NERC%20Interconnections.pdf.
[6] Denholm, Paul, et al., *Inertia and the Power Grid: A Guide without the Spin,* Golden, CO: National Renewable Energy Laboratory , 2020.
[7] North American Electric Reliability Corporation, "Reliability guideline primary frequency control." 2019.

[8] Independent Electricity System Operator, "Market rules chapter 4 grid connection requirements – Appendix 4.2." 2021.

[9] Independent Electricity System Operator, "Market manual 1: Connecting to Ontario's power system Part 1.6: Performance validation." 2021.

[10] ENTSO-E, "Fast frequency reserve – Solution to the Nordic inertia challenge." 2019, [Online]. Available: https://www.statnett.no/globalassets/for-aktorer-i-kraftsystemet/utvikling-av-kraftsystemet/nordisk-frekvensstabilitet/ffr-stakeholder-report_13122019.pdf.

[11] Australian Energy Market Commission, "Feasibility of fast frequency response obligations of new generators." 2017.

[12] North American Electric Reliability Corporation Reliability Guideline, "BPS-connected inverter-based resource performance" 2018.

[13] ENTSO-E, "Technical requirements for fast frequency reserve provision in the Nordic synchronous area – External document." 2021. [Online]. Available: https://www.statnett.no/globalassets/for-aktorer-i-kraftsystemet/marked/reservemarkeder/ffr/technical-requirements-for-ffr-v1.1.pdf.

[14] Roscoe, A., Knueppel, T., Da Silva, R., Brogan, P., Gutierrez, I., Elliott, D., & Perez Campion, J. C. (2020). "Response of a grid forming wind farm to system events, and the impact of external and internal damping." *IET Renewable Power Generation*, 14(19), 3908–3917.

[15] Keller, J. & Kroposki, B., *Understanding Fault Characteristics of Inverter-Based Distributed Energy Resources*, Golden, CO: National Renewable Energy Laboratory, 2010.

[16] National Grid ESO, "GC0111: Fast fault current injection specification text", 2020.

[17] Gurule, N. S., et al. "Grid-forming inverter experimental testing of fault current contributions." *2019 IEEE 46th Photovoltaic Specialists Conference (PVSC)*. Chicago, IL: IEEE, , 2019.

[18] NERC Reliability Standard PRC-024-2, "Generator frequency and voltage protective relay settings", 2015.

[19] Congress, "Energy policy act of 2005." Sec. 1254 Interconnection, 2005.

[20] Institute of Electrical and Electronics Engineers, IEEE Std, *IEEE Standard for Interconnection and Interoperability of Distributed Energy Resources with Associated Electric Power Systems Interfaces*. San Diego, CA: IEEE. pp. 1547–2018, 2018.

[21] Ellis, A., et al. "Review of existing reactive power requirements for variable generation." *2012 IEEE Power and Energy Society General Meeting*. San Diego, CA: IEEE, 2012.

[22] Ellis, A., et al. *Reactive Power Interconnection Requirements for PV and Wind Plants – Recommendations to NERC*. Albuquerque, NM: Sandia National Laboratories, 2012.

[23] California Independent System Operator Corporation, "Fifth replacement tariff appendix D black start generating units." 2017.

[24] ScottishPower Renewables, *Black-Start Capability - A Global first for ScottishPower Renewables*, [Online]. Available: https://www.scottishpowerrenewables.com/pages/innovation.aspx.

[25] Sørensen, T. B., Kwon, J. B., Jørgensen, J. M., Bansal, G., & Lundberg, P. "A live black start test of a HVAC network using soft start capability of a voltage source HVDC converter." Cigre Symposium Aalborg, 2019.

4 Toward Performance-Based Requirements and Generic Models for Grid-Forming Inverters

Deepak Ramasubramanian, Wenzong Wang,
Evangelos Farantatos, and Mohammad Huque
Electric Power Research Institute

CONTENTS

4.1 INTRODUCTION

The increase in inverter-based resources (IBRs) in transmission and distribution systems around the world, driven by government renewable energy policies and motivation to address climate challenges, is changing the dynamic behavioral characteristics of the power network. This increase in IBRs is accompanied by a reduction in the active power-generating synchronous machine fleet [1]. A consequence of this reduction in the fleet of rotating machines with mechanical inertia is a reduction in both transient reactive power support and short-circuit current delivery. This reduction in short-circuit current delivery is also known as reduction in short-circuit strength or stiffness of the network [2]. A reduction in system stiffness results in increased voltage sensitivity to change in current injections, both in magnitude and phase. This increase in voltage sensitivity can subsequently result in reduction in system damping and synchronization capability of active elements (such as rotating machines and IBRs) [3].

DOI: 10.1201/9781003302520-4

Conventional IBR plants generally have control loops with high bandwidth (i.e., with low rise time resulting in fast control loops), which enable them to maintain tight control of the active and reactive power injected by the IBR device into the grid. These control loops, if not tuned appropriately, can suffer from instability under reduced system stiffness [4]. This behavior of IBR control architectures has given rise to concerns regarding the stability of the future power system with increased percentage of IBRs.

Traditionally, low short-circuit strength scenarios are related to large values of Thévenin impedances (electrically far away), when viewed from a certain bus in the power system. However, with an increase in IBRs and subsequent reduction in synchronous machine fleet, a low short-circuit scenario could also manifest in electrically close networks simply due to the reduction in size and number of synchronous machines in the network. The dynamic behavior of such a network can be different from the dynamic behavior of a network with large Thévenin impedance values. As a result of this difference in types of low short-circuit networks, different IBR control techniques can have varied limitations for stable operation.

Although there exist many island power networks or microgrids that can run with only IBRs [5], the extension of this concept to large power networks with IBRs from different vendors gives rise to a circular problem [6]. As IBR percentage increases, IBRs would be expected to provide a wide variety of system services through their dynamic behavior [7]. At present, there are many different types of emerging IBR controls being designed and developed by researchers and inverter vendors [8–10]. A primary objective of these emerging control topologies is to be superior to existing IBR control topologies (which may suffer from instability with reduced voltage stiffness) while providing the additional services required by the IBR-dominated power system. To have an efficient design of these emerging control techniques, exact performance requirements must be known, which can only be specified either through standards and/or interconnection requirements from power system planners. However, writing either standards or interconnection requirements requires an idea of the capabilities of the future IBR control architecture.

For a future power system, it is important that system planners do not enforce requirements based on any specific types of grid-forming inverter control. This can result in restriction of performance and products that may enter the marketplace. Further, from a control system point of view, there are multiple different ways in which the same objective can be met. As a result, a control structure that might at first appear to be insufficient could possibly be tuned to be stable in low short-circuit scenarios. To accommodate these scenarios, the application of system performance requirements can be more flexible than exact control specifications [11].

In this chapter, it will first be shown that even conventional IBR control schemes can be designed and tuned to operate well in low short-circuit systems. The goal here is not to show that these modified controls may be stable in all scenarios or networks, but rather to show that suppose they can be tuned to be stable for few scenarios, it indicates the possibility of applying the particular control scheme, which a manufacturer can then tune for different scenarios. As a result of this exercise, fundamental nuances regarding grid-forming behavior can be uncovered which can then lead to the development of draft interconnection requirements that can be applied across systems for new resources. These interconnection requirements are subsequently based on the services and functions that the IBR would be expected to provide, rather than the type of control. In terms of dynamic performance requirements, they can be developed based on input-output time domain performance of the IBR, which can be achieved by the inverter manufacturer through different grid-forming inverter control architectures of their choice.

The layout of the chapter is explained as follows: first it is shown that despite the type of grid-forming control that is adopted, operational similarities exist across the various types of control methods. As a result, a system planner and operator do not need to know the exact type of grid-forming controller that is to be included in the inverter. Rather, they can develop functional requirements and performance specifications as shown in Section 4.3 of the chapter. Any inverter manufacturer can then use these performance requirements as a guideline to develop and tune their controllers. Finally, in order to develop performance requirements for a future network, a generic model such as the one described in Section 4.4 of the chapter can be used.

4.2 OPERATIONAL SIMILARITY ACROSS DIFFERENT GRID-FORMING CONTROL TYPES

In this work, the operational behavior of four different types of grid-forming architecture (shown at a high level in Figure 4.1) will be explored. These four types of grid-forming methods are reasonably popular and upon which there are numerous recent research articles.

A virtual synchronous machine (VSM) form of grid-forming control structure mimics the swing equation of a synchronous machine to derive a relationship between active electrical power output and angular frequency [8]. The Type A droop form of grid-forming control structure utilizes a standard set of droop equations for a relationship between active power-frequency and reactive power-voltage. However, the input-output relationship within the droop equations is opposite to conventional droop equations. The inputs are active power and reactive power, while the outputs of the droop equations are angular frequency and voltage magnitude, respectively [12].

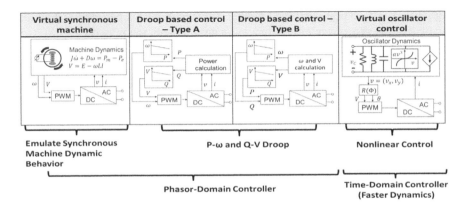

FIGURE 4.1 Four different types of grid-forming architecture considered for comparison.

A virtual oscillator type of grid-forming control structure utilizes properties of weakly coupled natural non-linear oscillators to set up limit cycle behavior in the voltage that is generated. Through appropriate parameterization, a pure sinusoidal wave can be constructed [13].

While these three forms of grid-forming control structures may appear to be different from each other, it has been shown recently [14] that at a basic equation level, a structural similarity exists between the VSM, Type A droop, and virtual oscillator (dispatchable, dVOC) methods of grid-forming control. This structural similarity allows for the same set of equations, with appropriate and suitable parameterization, to be used to represent either of these three types of grid-forming control. Subsequently, as the equation structure remains largely the same, the operational performance across all three methods of control also remains similar.

To this mix, one can now bring in the operational performance of the Type B droop-based method. The Type B droop-based control is implemented [15] with a fast voltage control loop and active power/frequency control loop to generate current commands. Here, the input-output relationship of the droop equations remains similar to the standard conventional droop equations with angular frequency and voltage magnitudes as inputs and active power and reactive power as outputs. Further, an inner current control loop and a phase-locked loop (PLL) are used to control current and synchronize the inverter resource to the grid.

Next, details about how the Type B droop-based control design (despite its current source control nature and the use of a PLL) can have operational performance similar to the other grid-forming controls are described. It is to be noted that the objective here is not to compare the performance of each grid-forming control against one another. Instead, the objective is to showcase that there is operational similarity across the various types of control.

4.2.1 TYPE B DROOP-BASED GRID-FORMING CONTROL

When differentiating between grid-forming inverters and grid-following inverters, a common distinction that has risen across the industry is the perception of the concept that a grid-following inverter is current-controlled source, while a grid-forming inverter is a voltage-controlled source. As a result of this perception, with the increase in the percentage of inverters in a power system, the presence of inverters behaving as current sources in a network is starting to be discouraged.

Here, however, it can be beneficial to go back to fundamental circuit theory and identify the nuances involved in current source versus voltage source operation of networks. Consider a simple network as shown in Figure 4.2. In this network, let us assume that all sources are current sources, while the load is represented as constant impedance.

Taking an arbitrary initial loading condition of $P_{ld1} = 100MW$ and $Q_{ld1} = 20MVar$, with network impedances of $Z_1 = 0.005 + j0.2$ pu and $Z_2 = 0.005 + j1.0$ pu on the system base of 100 Megavolt Amperes (MVA), values of voltage and current around the network can be evaluated. Now, if we go with the narrative that conventional inverters are current sources, then let us assume that the inverter denoted by the current source I_1 is operating at an output of $I_1 = 0.382 + j0.3969$ pu, while the inverter denoted by

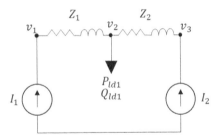

FIGURE 4.2 Simple electrical network to uncover nuance of current source operation in power systems.

the current source I_2 is operating at an output of $I_2 = -0.0128 + j0.5258$ *pu*. At this operating condition, with load represented as a constant impedance, using Kirchhoff's laws, it is fundamentally possible to evaluate the voltages around the network as,

$$v_1 = I_1(Z_1 + Z_{ld}) + I_2 Z_{ld},$$

$$v_2 = v_1 - I_1 Z_1, \qquad\qquad (4.1)$$

$$v_3 = I_1 Z_{ld} + I_2(Z_2 + Z_{ld}),$$

where Z_{ld} represents the impedance of the load evaluated at nominal voltage. Upon solving the equations, one obtains $v_1 = 1.045\angle84.51°$, $v_2 = 0.9763\angle79.54°$, and $v_3 = 1.012\angle110.262°$. This steady-state solution can be verified from an electromagnetic transient (EMT) simulation setup of this simple network. In the EMT network setup, to inject current appropriately, the current sources I_1 and I_2 use a simple synchronous reference frame (SRF) PLL to obtain the angle of the terminal voltage.

Therefore, from the basic circuit theory, it can be inferred that a circuit with only current sources can indeed achieve a steady-state solution. So now, if a conventional grid-following resource is perceived as a current source, then why there is a concern regarding increase in the number of conventional inverter resources in the power system? Does the power system require voltage sources and cannot operate with only current sources? Or can a 100% current source network continue to operate (again not considering a blackstart scenario) and generate voltages in the network based on the available impedances? Going back to the simple network, the answer lies in how the values of the current sources I_1 and I_2 are determined and more importantly how they are controlled when disturbances occur in the network.

Let us now consider a 5% increase in active power load at a point in time. Upon occurrence of the increase in load, if the values of I_1 and I_2 do not change fast (or don't change at all), then there is a definite possibility of frequency and/or voltage instability. However, occurrence of this instability is dependent on the type of load characteristic increase that occurs in this network. For example, with an increase in only active power load, even with no change in values of current injection, it is possible to achieve a stable voltage magnitude in the network as the load is assumed to have a constant impedance characteristic. However, frequency in the network will not be able to achieve a stable steady-state level as shown in Figure 4.3. Even if both

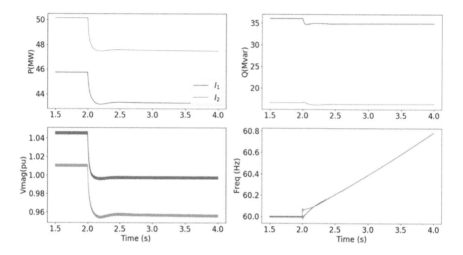

FIGURE 4.3 Response of simple network with only current sources for 5% load increase but with constant current injection.

active power and reactive power load change, with a constant impedance load characteristic, it would be able to achieve a steady-state voltage magnitude and root mean square (RMS) active and reactive power output; however, frequency will still not be able to achieve a stable value.

This behavior can be explained as follows. Although load has increased, frequency continues to increase as the load has a constant impedance characteristic. Due to the reduction in voltage magnitude, the power drawn by the load reduces. However, as the input active current has remained unchanged, there is surplus of active current being injected into the network. This results in an increase in frequency in the network.

This behavior is at the core of the concern related to an increase in grid-following inverters in a network. Almost all grid-following inverters in the network today are voltage source converters at their very core with fast inner current control loops and a PLL. However, just the presence of a PLL and a current source does not imply instability in a network with inverters. In addition to the presence of a current controller and a PLL, a key control feature of grid-following inverters is their operation in constant active power and reactive power mode [15]. This operation mode is the root cause of instability in grid-following inverters. In some scenarios, even if grid-following inverters do provide variation in active power and reactive power injection (also called as grid-supporting features), the time constant of change is usually of the order of seconds, which can be insufficient to maintain stability in a network.

To achieve stability in a network with a large percentage of inverters, an inverter that can change its current injection in a fast manner is required to maintain stability. Additionally, this fast change in current should be along both active power and reactive power axis. If such a change in current can be brought about, then even with a PLL and a current source, both voltage and frequency can be maintained in a stable manner in a network with large percentage of inverters.

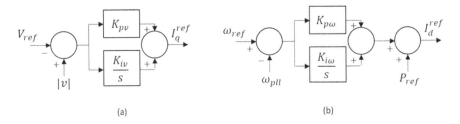

(a)　　　　　　　　　　　　　　(b)

FIGURE 4.4 Structure of controllers to generate reactive and active current commands to enable stable operation in the simple electrical network.

This concept can be validated using the same simple electrical network. For each current source, reactive current command/injection can be defined as the output of a controller whose objective is to maintain terminal voltage magnitude toward a fixed reference value, while active current can be defined as the output of a controller whose objective is to control frequency toward a fixed reference value. The generation of both current commands can be derived as shown in Figure 4.4. Here the individual current source's controller is structured in the dq frame oriented with the main network axis at the angle provided by a PLL. Further, it is assumed that the q axis leads the d axis.

For the same load increase of 5% in active power, with $K_{pv} = 1.0$, $K_{iv} = 20.0$, $K_{p\omega} = 0.33$, $K_{i\omega} = 2.0$, the response of the network is now as shown in Figure 4.5. The ability of the current sources to quickly change their values of injected current to a suitable value based upon the observed reduction of terminal voltage magnitude and electrical frequency (obtained from the PLL) enables the network to achieve a stable post disturbance operation point.

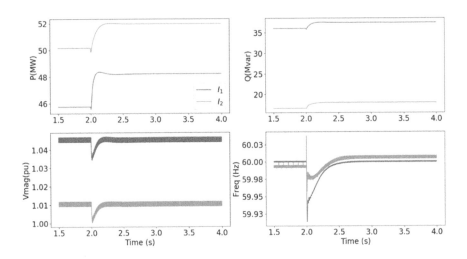

FIGURE 4.5 Response of simple network with only current sources for 5% load increase but with controlled current injection.

It can hence be inferred from here that it is possible to continue to operate an electrical network with only current sources, provided these currents can change their values of injected current in a reasonably fast manner when a disturbance occurs. The exact speed and magnitude of change that is required can subsequently be dictated by interconnection requirements laid out by the power system planner and operator.

The exercise carried out with this simple example allows for the exploration of similarities across various grid-forming control methods that have been proposed in research literature, with some variations also being used by inverter manufacturers. If an operational similarity exists across these various grid-forming methods, then it would be beneficial for power system planners and operators as they now may not need to specify exact control methodology for future inverter but can rather specify a set of performance requirements that are to be met by the device. The design and tuning of the exact control architecture that is needed to meet these performance requirements is then left to the inverter manufacturer.

An important point to highlight here is blackstart of a network or no-load open-circuit operation of a network is not being considered. A distinction is made between grid-forming behavior for blackstart operation and grid-forming behavior for continued operation of a network once the last synchronous machine trips.

4.2.2 GRID-FORMING CONTROL COMPARISON

To demonstrate the operational performance similarities of the behavior across four grid-forming control methods (VSM, Type A droop, Type B droop, dVOC), a full pulse width-modulated (PWM) switching model of a transmission-connected 70 MVA solar plant was constructed in an EMT simulation program. The solar plant was connected to an infinite source through an impedance $R_{grid} + jX_{grid}$ as shown in Figure 4.6. For this example, it is assumed that the solar plant does not have a plant controller as the inverters within the plant are in grid-forming mode. The details of the implementation of each of the four types of grid-forming methods are provided in references [14,15].

The value of $R_{grid} + jX_{grid}$ can be set based on a required value of short-circuit ratio (SCR) at the network interface of the IBR plant, for a given value of plant capacity/rating. In this comparison, an SCR of 1.5 is chosen with the solar plant dispatched at an active power set-point of 60 MW and a voltage reference set-point of 1.017 pu. While a lower value of SCR could be considered, care must be taken in these situations as the value of $R_{grid} + jX_{grid}$ could become unrealistic. For understanding the behavior of IBRs at lower values of SCR, it may instead be more accurate to construct

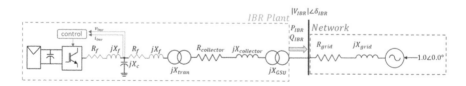

FIGURE 4.6 Single inverter infinite bus network to compare operational characteristics of grid-forming methods.

a small system with load and other dynamic devices like synchronous machines, and then bring about low values of SCR through the trip of synchronous machines. This approach will be expanded upon later in this chapter in the next section.

To bring about a disturbance in this single inverter infinite bus network, a 10% step change (both increase and decrease) in the magnitude of the grid voltage is applied. As the boundary bus with the network would be the point of applicability for a system planner/operator, the response of the plant at this location is observed as shown in Figure 4.7. The measured voltage is filtered through a 20 ms time constant to evaluate the RMS value for a system frequency of 60 Hz. It can be observed that the closed-loop response of all four control methods is operationally similar both in characteristic and from a time constant perspective. The closed-loop settling time of the voltage control loop can be observed to be 1.0 s, which results in a time constant of around 250–300 ms for the control loop. Such a definition of settling time and time constant can be used by system planners/operators to define future inverter performance characteristics.

The comparison of the operational behavior across all four control methods for a change in grid frequency is shown in Figure 4.8. Here, a 0.2 Hz frequency change (both increase and decrease) is applied at a rate of change of 10 Hz/s. Again, the operational behavior is seen to be similar across all four methods of grid-forming control with a closed-loop settling time of 1.0 s in the active power control loop.

The operational behavior of different grid-forming inverter control methods has also been compared on a distribution feeder [16,17] and similar behaviors such as fast reactive power response following a voltage disturbance have been observed. These behaviors are beneficial to the operation of weak distribution systems and can be required by system planners/operators if certain distribution systems become weaker in the future.

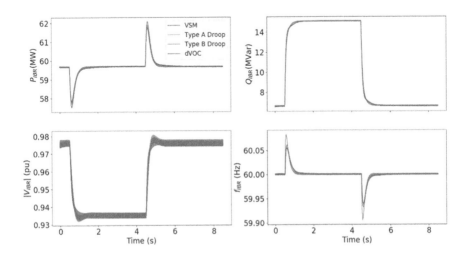

FIGURE 4.7 Comparison of dynamic response in EMT domain of four grid-forming methods to a 10% step change in grid voltage magnitude.

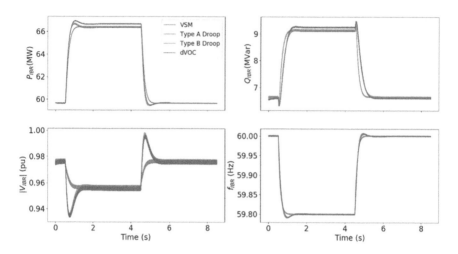

FIGURE 4.8 Comparison of dynamic response in EMT domain of four grid-forming methods to a 0.2 Hz change in grid frequency with 10 Hz/s rate of change.

The presence of this operational similarity now opens up the possibility of a power system having multiple different grid-forming control methods co-exist in a reasonable manner. Of course, there can be possibilities of control interactions among the various inverters and other dynamic devices; however, that is a concern even in today's power network which can be tackled using a variety of methods and techniques, which could also include technical minimum capability requirements in equipment standards.

4.3 TOWARD PERFORMANCE-BASED INTERCONNECTION REQUIREMENTS FOR GRID-FORMING INVERTERS

As grid-forming behavior can be derived and obtained in numerous ways as shown in the previous section, it can be beneficial for the power system industry to avoid placing restrictions on the type of, and manner in which, control elements could be used by an inverter manufacturer. Control agnostic definitions of performance and services are hence important to be defined. An example of control agnostic performance requirements are those discussed in Section 4.2 of this chapter regarding closed-loop settling time for step changes in grid voltage and frequency, at a low value of short-circuit strength. It has been shown that with a closed-loop settling time of around 1 s, the grid-forming inverter is able to bring about stable operation of a 100% inverter system upon the trip of the last synchronous machine. It should be noted here that performance specifications during blackstart and system restoration could go over and beyond the specifications for normal system operation. Blackstart and system restoration is a special operation mode even in today's power system paradigm. The focus for majority of the specifications would be to maintain continued stable and reliable operation of the power system even with the trip of the last synchronous machine (generator or condenser).

There are few bulk power system operators around the world that have drafted preliminary technical requirements for advanced inverter capability. National Grid Electric System Operator (NGESO) was one of the first system operators to begin to prepare a draft of technical specifications that future inverters would have to achieve. These specifications were complimented through the release of competitive tenders, called Stability Pathfinders, which enabled procurement of systems services, both old and new, from inverters. The development of these technical specifications from NGESO was initially through the Virtual Synchronous Machine Expert Group [18] which later became the Great Britain Grid Forming Working Group. This resulted in the development of a proposed grid code change (GC0137) [19]. The primary services to be procured include services to constrain the rate of change of frequency (RoCoF) both at the national level and local level, and services to maintain a suitable voltage following a fault along with the ability to ride through phase jumps. Synchronous resources can also be tagged as grid-forming resources here, provided they have the appropriate control systems available.

According to GC0137, a source is considered to be grid forming if it contains a synchronous internal voltage behind a real impedance, which in many ways is like a synchronous machine. Further, the main grid-forming capability features identified are along the lines of those that have usually been provided by synchronous machines. Some among these are contribution to synchronizing torque, instantaneous injection of fast fault current at the instant of fault occurrence (both due to voltage magnitude and phase change), and contribution to damping torque. Per the proposed grid code change, a grid-forming resource is also expected to be able to continue to inject active power, without entering current limited operation, even upon the occurrence of a phase jump as large as 60°. There is also an expectation of capability to operate at a zero short-circuit level.

In the United States, recently in 2021–2022, the Hawaiian Electric Company (HECO) requested grid-forming capability from newly proposed projects that included both stand-alone storage and hybrid PV-storage installations [20]. Key aspects related to this requested grid-forming capability include the ability to support system operation without requiring the presence of synchronous generation resources and continue to operate in a stable and reliable manner upon the trip of the last synchronous machine (loss of grid reference). Further, switching between a grid-forming mode and any other mode of operation is not advised. A key point in these requirements is that grid-forming resources are to secure adequate energy storage to contribute to system restoration, if asked by the system operator. Additionally, short-term overcurrent capability may be required during system restoration to energize transformers, lines, and motors, along with the provision of a ground reference path.

In Australia, the Australian Electricity Market Operator (AEMO) has taken a gradual (stepwise) approach to obtain various services from advanced inverters [21]. These capabilities and services are expected to increase with the increase in the percentage of inverters in the system. AEMO's approach to grid-forming inverters is based upon an application perspective with a near-future application being provision of stabilizing services for inverters in weak grids. Subsequent applications include contributing to system security, stable islanded operation, and system restoration/restart.

AEMO and HECO recommend prioritizing the use of grid-forming capability on energy storage elements such as batteries due to a large number of energy storage projects in the interconnection queue along with the availability of an energy buffer.

A common thread running through these various emerging requirements for grid-forming inverters are performance capabilities along the lines of [22,23]:

1. Ability to create an open-circuit voltage at the point of interconnection of the inverter plant. Note that this characteristic is defined at the plant point of interconnection and not at the terminal of an individual inverter within the plant. This implicitly assumes that the inverter plant is capable of,
   a. Serving its own auxiliary load, and
   b. Operating in the absence of a synchronous machine.
2. Ability to synchronize and operate in conjunction with other sources of energy in the grid. These other sources can include conventional rotating machines as well as other forms of IBR control methods. This also includes different types of load.
3. Upon the occurrence of a large load/generation event, to have the ability to contribute toward arresting the increase/decline of frequency and also contribute toward the subsequent return of frequency to the nominal value.
   a. The contribution of the inverter should be such that it shares the burden with other participating resources.
4. Contribute toward provision of reactive power support and voltage regulation within the continuous operation region. For abnormal grid conditions, contribute toward aiding fast and stable voltage recovery.
5. If required by protection relay algorithms, contribute requisite level and type of fault current. The implicit acknowledgment here is the understanding that provision of fault current above a certain threshold might require changes/upgrades in the inverter hardware.
6. At a minimum, to not negatively impact the damping of naturally occurring oscillations in the power system following major disturbances. Moreover, if the inverter is situated such that it has the capability to improve the damping of oscillatory mode to have the necessary supplemental controls to facilitate this when properly tuned and designed.
7. Contribute toward minimization of unbalanced operation (within the continuous region of operation) and improvement of power quality of the network.

Some of the abovementioned requirements are still rather high level and are evolving. Moreover, since they are developed mainly by bulk system operators, the focus is mostly on maintaining system stability with grid-forming inverters.

When it comes to microgrid operation, even though maintaining stability remains important (especially with multiple generation resources), the concern on power quality increases. Developing grid-forming inverter requirements has not been practiced by most of the distribution operators. However, microgrid field test procedures have been developed at a few utilities to ensure that the system-level power quality requirements can be met with the grid-forming plant.

Next, the considerations and procedure to develop control agnostic performance requirements for grid-forming plants in utility-level microgrid operation are discussed through an example. For single customer microgrids or small-scale dedicated microgrid with only a few customers, the power quality requirements may be negotiated among the parties involved. However, for utility-level microgrids that involve at least part of a medium voltage feeder and various customers at different locations, it is expected that the power quality should be maintained according to the established standards.

The generation resources inside the microgrid should therefore contribute to meeting these system-level power quality requirements. Here a simple case is studied where there is only one grid-forming plant and it is the only generation resource in the microgrid. Note it is common in today's microgrid design to have one large grid-forming plant serve the entire microgrid. In this scenario, the single grid-forming plant is responsible for maintaining the power quality of the system (with the help of other voltage regulating devices, if available), and its performance requirements should be specified to achieve this goal. Based on this rationale, the general procedure to establish the grid-forming plant requirements is to first identify the system-level power quality requirements that should be maintained and subsequently identify the services and functions of the grid-forming plant to contribute to meeting the system-level requirements through detailed studies.

To illustrate this procedure, voltage balancing is chosen as the microgrid system-level requirement under consideration. Since unbalanced load always exists on any distribution feeder and can be aggravated in microgrids, it is important that the voltage unbalance/imbalance on a three-phase feeder be controlled under certain limit. For instance, ANSI C84.1 recommends that the voltage unbalance be maintained within 3%. If the voltage unbalance is severe, it can lead to motor derating and potentially cause damage to three-phase induction motor loads. If the feeder is effectively grounded, the voltage unbalance will appear in the form of negative sequence voltage. Therefore, negative sequence voltage control is an important requirement for a grid-forming plant in microgrid operation.

Note that when a distributed energy resource (DER) is not working in a microgrid but in distribution grid-connected mode, interconnection standards such as IEEE Std 1547™-2018 do not require it to regulate the negative sequence voltage. Instead, it is assumed that the negative sequence voltage is regulated by the substation and the generation resources in the transmission system. However, this is no longer true in a microgrid, where the negative sequence voltage caused by unbalanced load must be controlled by the generation sources inside the microgrid.

To facilitate the development of detailed requirements on negative sequence voltage control of the grid-forming plant, EMT simulation studies have been conducted on a real utility-level microgrid circuit. The microgrid circuit is a section of a 12.47 kV feeder, which can be isolated and operate as a microgrid. The peak load of the microgrid is around 3 MW with an average power factor of 0.88 (absorbing reactive power). An energy storage grid-forming plant with 8,250 kVA is the only power source inside the microgrid, which is modeled in an EMT simulation software with both positive and negative sequence controls.

TABLE 4.1

Grid-Forming Plant Negative Sequence Control Case Studies

Case #	Negative Sequence Control Objective	Negative Sequence Current Capability
1	Regulate negative sequence current to zero	None
2	Regulate negative sequence voltage at PCC to zero	0.05 pu
3	Regulate negative sequence voltage at PCC to zero	0.1 pu

Three different cases are studied to investigate the impact of grid-forming plant negative sequence voltage control on system voltage unbalance, as shown in Table 4.1. The first case considered is where the grid-forming plant regulates its negative sequence current to zero and therefore does not regulate the negative sequence voltage. Note that without explicit requirement, an inverter may choose to eliminate its negative sequence current injection to reduce the DC ripple.

In cases 2 and 3, the grid-forming plant regulates the negative sequence voltage at its point of common coupling (PCC) to zero, when the negative sequence current output is within its capability/limit. Compared to case 2, in case 3, the grid-forming plant has higher negative sequence current capability, meaning it can inject higher negative sequence current (up to 0.1 pu) to regulate the negative sequence voltage.

Voltages at different locations across the three-phase feeder inside the microgrid have been simulated and the results are shown in Figure 4.9. The disturbance considered is disconnection of part of the feeder with loads. As a result of the disturbance, the level of load unbalance increases in the remainder of the microgrid. The shaded area in the figure corresponds to the normal operation voltage range defined in ANSI C84.1 range A. As can be seen from the simulation results, when the grid-forming plant is not regulating the negative sequence voltage (case 1), the voltage unbalance in the system is severe, which will cause damage to the three-phase motor loads in the microgrid. At the same time, the high negative sequence voltage also results in

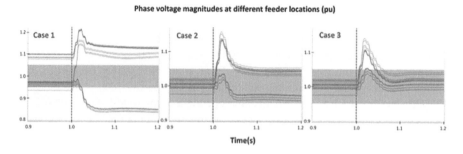

FIGURE 4.9   Voltage magnitude of each phase at different locations on the 12.47 kV feeder inside the microgrid.

individual phase voltage violations. In cases 2 and 3, when the grid-forming plant is regulating negative sequence voltage at its PCC, the voltage unbalance is reduced across the feeder and the individual phase voltages are within the ANSI C84.1 range A. Moreover, the voltage unbalance is less in case 3 with higher negative sequence current capability from the grid-forming plant.

Based on this study, it becomes clear that the grid-forming plant in microgrid operation should regulate the negative sequence voltage with sufficient negative sequence current capability. Therefore, a preliminary performance requirement for grid-forming plant can be developed as: "A grid forming power plant should maintain balanced voltage at its PCC when it operates within the negative sequence current capability and total current capability and the negative sequence current capability should be higher than x". Notice the negative sequence current capability x that should be required is dependent on the load characteristics of a specific microgrid and should be set forward based on detailed studies. As mentioned, this study considers the scenario where there is only one grid-forming plant inside the microgrid. When there are multiple grid-forming plants, the requirement should be adapted to ensure proper negative sequence current sharing among the grid-forming plants, while maintaining the voltage unbalance within a certain limit.

Other requirements for grid-forming plants in microgrid operation include voltage harmonics regulation, frequency and voltage regulation, maintaining stability, fault ride-through, and blackstart (may not be required for every grid-forming plant if there are multiple). These requirements are being actively developed [24]. Whether the same requirements can and should be applied to grid-forming plants under different interconnection and operation scenarios (e.g., microgrid, distribution connected, transmission connected) requires further research.

4.4 GENERIC MODEL DEVELOPMENT FOR LARGE SYSTEM STUDIES

The proprietary nature of inverter controls may at times restrict the ability to carry out long-term planning studies as adequate dynamics models for these resources may not be available. However, adequate and sufficient models are required by system planners to efficiently be able to carry out studies that can provide insight into need for additional assets in the system. For example, if a new high-voltage transmission corridor is to be constructed before the system inverter resource percentage level reaches a certain value, then the studies needed to obtain the necessary approvals to build the transmission corridor would have to be carried out quite a few years before the percentage level is obtained.

Additionally, due to the weaker nature of the changing power system, verification of operational performance of a resource in a single inverter infinite bus setup may not be sufficient to gain confidence in the ability of the resource to contribute toward improving the stability of the grid. Instead, larger system studies may have to be carried out, but this may bring about a computational burden on planners and operators.

To help address both challenges described above, generic grid-forming control models in positive sequence simulation software can be beneficial for transmission system planners and operators. Positive sequence simulation software has been used as the default dynamic simulation environment by transmission planners for the past couple of decades. These simulators were, however, designed from the perspective of synchronous machine electromechanical dynamics which are usually 0.1–10 Hz around nominal. Positive sequence simulators are computationally efficient and afford efficient simulation of large power networks. However, in such a simulation environment, there can be concern regarding the accuracy of representation of IBR dynamic behavior. Although it is understood that positive sequence simulators would not be able to represent all the dynamics associated with inverter resources, recent advances in the development of positive sequence models for inverters [25] in low short-circuit networks offer promise to transmission system planners and operators.

The concept that is used in the construction of these low short-circuit positive sequence inverter models can, however, be used to develop generic positive sequence models for grid-forming inverters. Building upon the operational similarity across different grid-forming methods, the structure of a generic positive sequence can be constructed as shown in Figure 4.10. Depending on the parameterization of the model, each of the different types of grid forming can be enabled. In-depth details related to the parameterization of this model are available [26].

To showcase the performance of this generic positive sequence model, a test network as shown in Figure 4.11 was constructed in both EMT domain and positive sequence domain using commercial simulation tools used widely in the power system industry. This network is a modification of the IEEE Standard 9 bus system.

The sources on the right side of the network at buses 3 and 14, respectively, represent synchronous sources with the source at bus 3 being a 150 MVA synchronous generator with a static excitation system and governor, while the source at bus 14

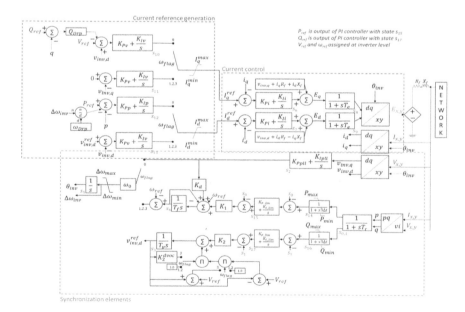

FIGURE 4.10 Structure of generic positive sequence grid-forming model to capture structural and operational similarity across four different grid-forming methods.

FIGURE 4.11 Performance of generic grid-forming positive sequence model, modified IEEE 9 bus network with conventional inverter resource at bus 13, synchronous generator at bus 3, and synchronous condenser at bus 14.

represents a 50 MVA synchronous condenser with a static excitation system. IBRs are connected to bus 7 through a 200 km long 230 kV transmission corridor. The source at bus 13 represents a 200 MVA conventional solar plant which has a plant controller that controls the voltage magnitude and active power output at bus 11 to fixed setpoints by providing active and reactive power commands downstream to the inverter control. A 100 ms communication delay is assumed to be present between the plant controller and inverter controls. In positive sequence domain, the dynamic behavior of this conventional inverter resource is represented using the Western Electricity Coordinating Council (WECC) generic models REGC_C + REEC_D + REPC_A [25,27,28]. The total load in the network is approximately 215 MW with constant current static characteristic for active power and constant impedance static characteristic for reactive power. The short-circuit capacity at bus 10 is 376 MVA. This configuration of the system is assumed to be the base case configuration. The dynamic response of the network for a 10% increase in active power load is shown in Figure 4.12.

It is observed that the trend of response across both EMT simulation domain and positive sequence simulation domain is similar. It is further acknowledged that although in this work the same values of control gains have been used in both the EMT model and the positive sequence model, due to inherent differences between the two simulation domains, some amount of tuning of the values of control parameters could be required to be carried out in the positive sequence simulation environment to obtain a better match of the responses [29]. Like the load increase event, the trend of response for a solid to ground 3 cycle three-phase fault at bus 15 is also comparable across the positive sequence domain simulation and EMT domain simulation as shown in Figure 4.13.

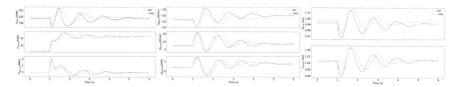

FIGURE 4.12   Dynamic response of modified IEEE 9 bus network in the base configuration with 10% increase in active power load.

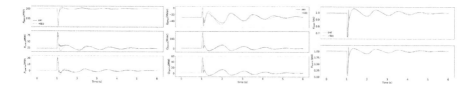

FIGURE 4.13   Dynamic response of modified IEEE 9 bus network in the base configuration for a 3 cycle solid to ground three-phase fault at bus 15.

To create an extremely weak scenario, both synchronous sources connected to bus 3 can be disconnected. It goes without saying that in the base configuration of the system, the dynamic response is unstable upon disconnection of both synchronous sources. Hence to this network, the 70 MVA grid-forming solar plant (whose operational behavior was discussed in the previous section) is connected at bus 10, but at an active power set-point of 0 MW to not vary the pre-disturbance steady-state power flow solution.

With the grid-forming inverter in the network, the operational response of all four grid-forming methods compared across EMT domain and positive sequence domain is shown in Figure 4.14 for the same 10% load increase event and the three-cycle solid to ground three-phase fault at bus 15. For both events, with the grid-forming inverter, the increase in damping in the network is immediately observable. Further, the response from the positive sequence model has a trend and trajectory that is similar to the response from the EMT model. These responses are encouraging as they showcase the ability of the generic positive sequence model to represent the trend of response that may be expected from grid-forming inverters, and hence, system planners and operators can carry out large system studies with high inverter percentage.

However, the takeaway from this section is not to indicate that with these newer positive sequence models, EMT studies do not need to be performed. These newer positive sequence models serve to bridge the gap between EMT and positive sequence simulators, thereby providing more tools in the transmission planner's/operator's toolkit. They still have room for improvement, and further development of these models is an area of continuous research.

An example of a potential limitation of the positive sequence model is shown in Figure 4.15. Here, the event applied is the simultaneous trip of both synchronous resources, thereby creating a 100% inverter system. With a 70 MVA rating and total load of 215 MW, the grid-forming device is approximately 30% of the entire load. The operational behavior of the two types of droop grid-forming modes is shown, across both positive sequence simulation domain and EMT simulation domain. Upon

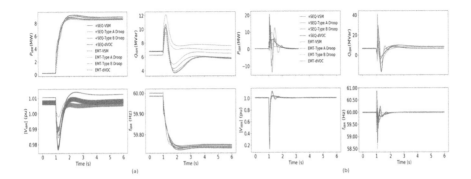

FIGURE 4.14 Dynamic response of modified IEEE 9 bus network with 70 MVA grid-forming inverter in four different control modes for (a) 10% increase in active power load and (b) a three-cycle solid to ground three-phase fault at bus 15.

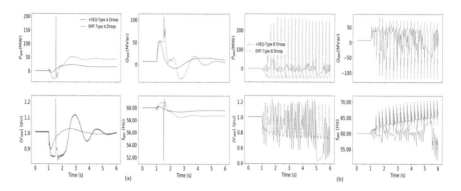

FIGURE 4.15 Dynamic response of modified IEEE 9 bus network with 70 MVA grid-forming inverter in (a) Type A and (b) Type B droop control mode for the trip of both synchronous sources, thereby creating a 100% inverter network.

first glance, one might make an inference that Type B grid-forming droop control has a potential limitation in operating in such a network, and this response is reflected in both EMT domain and positive sequence domain. Type A droop, however, appears to have a stable trajectory in positive sequence domain but has a bifurcation in EMT domain. The exact cause for this bifurcation is yet to be determined, and as a result, further modifications may also be necessary in positive sequence models.

However, if the rating of the grid-forming device is increased to 200 MVA, then the response of both types of droop control methods for this 100% inverter system event is shown in Figure 4.16. Here, both droop control methods are able to bring about a stable operation, albeit with a bit of an oscillatory response in Type B

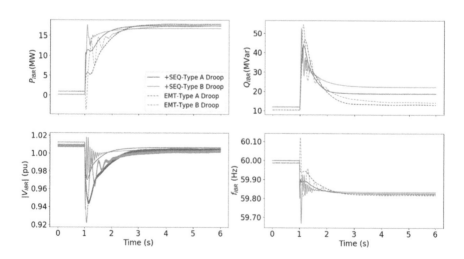

FIGURE 4.16 Dynamic response of modified IEEE 9 bus network with 200 MVA grid-forming inverter in Type A and Type B droop control mode for the trip of both synchronous sources, thereby creating a 100% inverter network.

droop method. Further, the trend of the response is also captured in positive sequence domain. Here again, the parameterization of the positive sequence model can benefit from a bit of fine tuning as is carried out in any common power plant parameterization exercise [29]; however, a system planner making use of these models will be able to obtain sufficient inferences regarding the operation of the network.

The development of this generic positive sequence model in simulation tools used widely in the power system industry to represent grid-forming inverter control methods provides a transmission planner and operator the ability to being the planning of a future power system. Further, it allows for the framing of control structure agnostic interconnection requirements that can be passed onto the plant developer and inverter manufacturer who can then design and tune their product to meet the desired performance. In this manner, the grid receives the services it needs [7] while allowing inverter manufacturers to innovate and develop different forms of control architectures without a restriction on the use of specific control elements.

4.5 CONCLUSION

As IBR penetration increases in the power network, the percentage of grid forming technology is also expected to increase. With a similar operational performance across many different types of grid-forming control structures, system planners and operators could specify performance requirements that could be met by a variety of control technology. Further, the use of appropriately parameterized generic models across multiple simulation domains can aid in planning a future power system. Further research work is required to fully understand the operation of grid-forming resources under faults, both balanced and unbalanced. Additionally, the ability of the grid-forming resources to energize transformers and the motor load is to be studied. Additionally, it is yet to be understood if grid-forming inverters would add any new modes of oscillation in a network.

REFERENCES

[1] https://www.theguardian.com/australia-news/2022/feb/17/australias-largest-coal-fired-power-station-eraring-to-close-in-2025-seven-years-early.
[2] Kundur, P., Balu, N.J., and Lauby, M.G. *Power System Stability and Control*, McGraw-Hill, New York, 1994.
[3] Larsen, E.V. and Swann, D.A. Applying power system stabilizers part I: General concepts. *IEEE Transactions on Power Apparatus and Systems*, PAS-100(6), pp. 3017–3024, 1981.
[4] Li, Y., Fan, L. and Miao, Z. Stability control for wind in weak grids. *IEEE Transactions on Sustainable Energy*, 10(4), pp. 2094–2103, 2018.
[5] https://www.esig.energy/event/g-pst-esig-webinar-series-survey-of-grid-forming-inverter-applications/.
[6] High Share of Inverter-Based Generation Task Force. *Grid-Forming Technology in Energy Systems Integration*. Energy Systems Integration Group, Reston, VA, 2022.
[7] IBR Research Team Services Group. *System Needs and Services for Systems with High IBR Penetration, Global Power Systems Transformation (G-PST) Consortium*, 2021. https://globalpst.org/wp-content/uploads/GPST-IBR-Research-Team-System-Services-and-Needs-for-High-IBR-Networks.pdf

[8] Beck, H.P. and Hesse, R. Virtual synchronous machine. In *2007 9th International Conference on Electrical Power Quality and Utilization*, pp. 1–6. IEEE, Barcelona, Spain, 2007.

[9] Du, W., Chen, Z., Schneider, K.P., Lasseter, R.H., Nandanoori, S.P., Tuffner, F.K. and Kundu, S. A comparative study of two widely used grid-forming droop controls on microgrid small-signal stability. *IEEE Journal of Emerging and Selected Topics in Power Electronics*, 8(2), pp.963–975, 2019.

[10] Dhople, S.V., Johnson, B.B. and Hamadeh, A.O. Virtual oscillator control for voltage source inverters. In *2013 51st Annual Allerton Conference on Communication, Control, and Computing (Allerton)*, pp. 1359–1363. IEEE, Monticello, IL, 2013.

[11] The Electric Power Research Institute. *Grid Forming Inverters: EPRI Tutorial*. EPRI, Palo Alto, CA, p. 3002021722, 2021.

[12] Chandorkar, M.C., Divan, D.M. and Adapa, R. Control of parallel connected inverters in standalone AC supply systems. *IEEE Transactions on Industry Applications*, 29(1), pp.136–143, 1993.

[13] Lu, M., Purba, V., Dhople, S. and Johnson, B. Comparison of droop control and virtual oscillator control realized by Andronov-HOPF dynamics. In *IECON 2020 The 46th Annual Conference of the IEEE Industrial Electronics Society*, pp. 4051–4056, IEEE, Singapore, 2020.

[14] Johnson, B., Roberts, T., Ajala, O., Dominguez-Garcia, A. D., Dhople, S., Ramasubramanian, D., Tuohy, A., Divan, D. and Kroposki, B. A generic primary-control model for grid-forming inverters: Towards interoperable operation & control. In *2022 55th Hawaii International Conference on System Sciences (HICSS)*, Maui, HI, USA, pp. 1–10, 2022.

[15] Ramasubramanian, D., Baker, W., Matevosyan, J., Pant, S. and Achilles, S. Asking for fast terminal voltage control in grid following plants could provide benefits of grid forming behavior. *IET Generation Transmission Distribution*, 00, pp. 1–16, 2022.

[16] Wang, W., Ramasubramanian, D., Kannan, A. and Strauss-Mincu, D. Benefit of fast reactive power response from inverters in weak distribution systems. In *2022 IEEE Rural Electric Power Conference*, Savannah, GA, USA, 2022.

[17] The Electric Power Research Institute. *Program on Technology Innovation: Benefit of Fast Reactive Power Response from Inverters in Supporting Stability of Weak Distribution Systems: A Use Case of Grid Forming Inverters and their Performance Requirements*. EPRI, Palo Alto, CA, p. 3002020197, 2020.

[18] https://www.nationalgrideso.com/codes/grid-code/meetings/vsm-expert-workshop.

[19] https://www.nationalgrideso.com/document/220511/download.

[20] https://www.hawaiianelectric.com/documents/clean_energy_hawaii/selling_power_to_the_utility/ competitive_bidding/ 20190822_final_stage_2_rfp_book_3.pdf.

[21] https://aemo.com.au/-/media/files/initiatives/engineering-framework/ 2021/application-of-advanced-grid-scale-inverters-in-the-nem.pdf.

[22] The Electric Power Research Institute. *Grid Forming Inverters: EPRI Tutorial*. EPRI, Palo Alto, CA, p. 3002018676, 2020.

[23] Ramasubramanian, D., Pourbeik, P., Farantatos, E. and Gaikwad, A. Simulation of 100% inverter-based resource grids with positive sequence modeling. *IEEE Electrification Magazine*, 9(2), pp. 62–71, 2021.

[24] The Electric Power Research Institute. *Performance Requirements for Grid Forming Inverter Based Power Plant in Microgrid Applications: First Edition*. EPRI, Palo Alto, CA, p. 3002020571, 2021.

[25] Ramasubramanian, D., Wang, W., Pourbeik, P., Farantatos, E., Gaikwad, A., Soni, S. and Chadliev, V. Positive sequence voltage source converter mathematical model for use in low short circuit systems. *IET Generation Transmission Distribution*, 14, pp. 87–97, 2020.

[26] The Electric Power Research Institute. *Generic Positive Sequence Domain Model of Grid Forming Inverter Based Resource* EPRI Palo Alto, CA, p. 3002021403, 2021.

[27] *Proposal for New Features for the Renewable Energy System Generic Models*, WECC REMWG Group, 2021, https://www.wecc.org/Reliability/Memo_RES_Modeling_Updates_111121_Rev22_Clean.pdf.

[28] *Model User Guide for Generic Renewable Energy System Models*, EPRI, Palo Alto, CA, p. 3002014083, 2018.

[29] *Power Plant Parameter* Derivation *(PPPD) v14.0*, EPRI, Palo Alto, CA, p. 3002022342, 2021.

5 An Overview of Modeling and Control of Grid-Forming Inverters

Sam Roozbehani
Khajeh Nasir Toosi University of Technology Branch

Reza Deihimi Kordkandi and
Mehrdad Tarafdar Hagh
University of Tabriz

CONTENTS

5.1 INTRODUCTION

Distributed generation (DG) is a source of renewable and non-renewable electric energy that is directly connected to the distribution network or the consumer [1]. These DGs are classified into inverter based and non-inverter based. In recent years, the penetration rate of inverter-based DGs has increased due to renewable energy development in the power grid and microgrids [2].

As a result of merging DGs, energy storage systems and loads, distribution networks have changed from passive to active mode. The integration of traditional and renewable sources along with the corresponding electric loads on the side of the distribution network has introduced a structure called microgrid, which should be capable of operating in both grid-connected and islanded modes. One of the basic issues of microgrids is power control and power sharing of the renewable and non-renewable inverter-based DGs to ensure stability of microgrid voltage and frequency.

The control mode of the inverter-based DGs has two layers, which includes inner and outer control layers. The main functions of the outer loop are power synchronization and voltage profile management. Accordingly, the control mode of the inverter-based DGs is classified into grid-forming and grid-following modes.

The grid-following units are controlled in power (current) control mode to generate active and reactive power based on the synchronized phase from grid so that to follow the grid. The grid-forming units are usually controlled in voltage control mode to generate the grid frequency and voltage so that to "form" grid. Most of the renewable-based DGs operate in this mode.

When microgrid operates in grid-connected mode, the inverter-based DGs usually operate in grid-following mode. In this condition, the synchronous generators regulate the voltage and frequency of the main grid. In contrast, when the microgrid is islanded, at least one of the inverter-based DGs must regulate the voltage and frequency of the microgrid. Therefore, the mentioned inverter should operate in grid-forming mode. Furthermore, when the microgrid is connected to weak grid, the grid-forming converters can play an important role to regulate voltage and frequency of the main grid [3], which is necessary for stability and flexibility of the islanded microgrids. In recent years, the use of grid-forming converters has increased as a result of the increase in the penetration rate of renewable inverter-based DGs. Therefore, the grid-forming converters have been introduced for grid-connected and islanded microgrids in the last decade.

Also, the inner loop is developed for calculating the modulation signal for regulating the output voltage of the inverter. Most of the publications described common cascaded proportional-integral (PI) controller for both voltage and current control whenever the inverter-based DGs work in grid-forming or grid-following modes. This type of control has two main drawbacks. One of them is slow response and long voltage recovery time after load steps and another is sensitive to parameter detuning.

Sliding mode control is one of the solutions for decreasing the sensitivity to parameter uncertainties and known for strong robustness in face of perturbations [4–7].

Most of the recent publications have used the current error as sliding surfaces and PI voltage controller in common cascaded form. In this condition, the problem of long recovery time in a load disturbance still exists. The authors in [8] have proposed the direct voltage control for solving long recovery time.

In order to overcome the issues mentioned in above paragraphs, in this chapter, the cascaded and direct voltage control strategies for grid-forming converters are developed via sliding mode and fractional-order sliding mode controllers to demonstrate their superiority over the conventional approaches. Also, conceptual explanation, differences of the grid-forming and grid-following converters, and cascaded and direct voltage controls are presented in the following sections.

5.2 BASIC DEFINITION OF GRID-FORMING CONVERTERS

According to different control objectives, inverter-based DG units are categorized as grid-forming and grid-following units [9]. In the former category, the voltage control mode is utilized to "form" a grid, or in other words, to produce the voltage and frequency of the grid.

The simplified system configuration of voltage control mode with direct droop is depicted in Figure 5.1a. With only inner loop of voltage control, the constant bus frequency and voltage can be fixed through operating the system as an ideal grid-forming unit. The additional outer loop based on direct droop control makes the systems be capable of regulating active and reactive powers. This subject is shown in the following equations:

$$f_{ref,i} = f_{rated} - m_{p,i} \cdot \left(P_i - P_{rated,i} \right) = f^* - m_{p,i} P_i, m_{p,i} = \frac{|\Delta f|}{P_{rated,i}} \tag{5.1}$$

$$V_{ref,i} = V_{rated} - n_{Q,i} \cdot \left(Q_i - Q_{rated,i} \right) = V^* - n_{Q,i} Q_i . n_{Q,i} = \frac{|\Delta V|}{Q_{rated,i}} \tag{5.2}$$

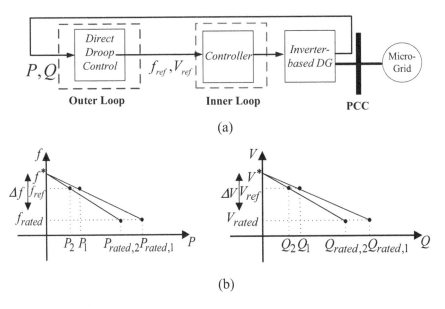

(a)

(b)

FIGURE 5.1 Simplified system configuration of grid-forming converter-voltage control mode [10].

In the above equations, "i" is the representative of each DG units, "V_{rated}" and "f_{rated}" correspond to nominal voltage and frequency of the microgrid, "Qi" and "Pi" represent the respective output reactive power and output active power of the DG, and "$Q_{rated,\,i}$" and "$P_{rated,\,i}$" denote the rated reactive power and rated active power of the DG. In addition, "$m_{P,\,i}$" and "$n_{Q,\,i}$" represent the droop slopes for active and reactive parts, and "$V_{ref,\,i}$" and "$f_{ref,\,i}$" illustrate the reference voltage and reference frequency of the DG, respectively. The selection of "$m_{P,\,i}$" and "$n_{Q,\,i}$" is of great significance regarding its effect on the network stability. In general, it is tried to coordinate the droops in a way that each DG system supply becomes proportional to its capacity.

The diagrams for $P-f$ and $Q-V$ droops of two DG units are presented in Figure 5.1b. The $Q_{rated,1}$, $Q_{rated,2}$ and $P_{rated,1}$, $P_{rated,2}$ values in the diagrams denote the rated reactive and active powers of two DGs. As can be seen in these diagrams, the active and reactive powers (P_1, P_2, Q_1, and Q_2) are generated by DGs at the reference frequency (f_{ref}) and voltage (V_{ref}) of the microgrid.

The second category of DGs, known as the grid-following units, is mainly controlled in power control mode. In this category, the active and reactive powers are generated on the basis of synchronized phase from grid. The simplified system configuration of power control mode with reverse droop is presented in Figure 5.2a. Constant power can be delivered as an ideal grid-following unit, with only inner loop of power control mode. The outer loop, which is on the basis of reverse droop control, provides the capability of bus voltage and frequency regulations. This subject is shown in the following equations [11]:

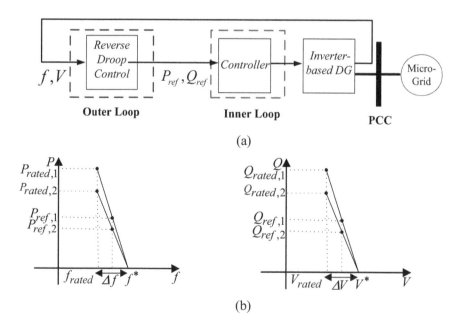

FIGURE 5.2 Simplified system configuration of power control mode-current control mode [10].

$$P_{ref,i} = P_{rated,i} - \frac{1}{m_{p,i}} \cdot \left(f_i - f_{rated} \right) = \frac{1}{m_{p,i}} f^* - \frac{1}{m_{p,i}} f_i \qquad (5.3)$$

$$Q_{ref,i} = Q_{rated,i} - \frac{1}{n_{Q,i}} \cdot \left(V_i - V_{rated} \right) = \frac{1}{n_{Q,i}} V^* - \frac{1}{n_{Q,i}} V_i \qquad (5.4)$$

In the above equations, "V_i" and "f_i" represent the measured voltage amplitude and frequency of each DG units, and "$Q_{ref,i}$" and "$P_{ref,i}$" correspond to the set-points of reactive and active powers, respectively. The droop diagrams for f-P, P-f, V-Q, and Q-V of two DG units can be seen in Figure 5.2b.

It is worth mentioning that the reverse droop approach is utilized in most of the non-dispatchable DGs, including wind power and photovoltaic systems. In the power control mode, the main role of the grid-supporting power converter is regulating the changes in the frequency and voltage of microgrid. In this regard, the variations in the frequency and amplitude of the voltage are controlled by the active and reactive powers, respectively.

5.3 MODELING OF GRID-FORMING CONVERTERS

The circuit model of the grid-forming converter consists of a DC voltage source representing the power supply connected to it, and an output inductive-capacitive (LC) filter to filter out the switching harmonic [12]. According to the controller, the state-space equations for modeling the converter can be obtained both in synchronous and stationary reference frames, which is discussed in the following subsections. This model is illustrated in Figure 5.3.

5.4 CONTROL OF GRID-FORMING CONVERTERS

5.4.1 GENERAL CONTROL STRUCTURE OF GRID-FORMING CONVERTERS

In this section, the purpose is to review grid-forming converter control methods. The comprehensive control structure for controlling a grid-forming converter is shown in Figure 5.4 [13].

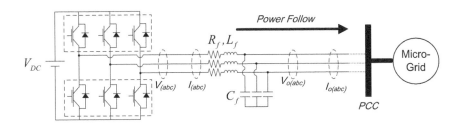

FIGURE 5.3 Circuit model of the grid-forming converter.

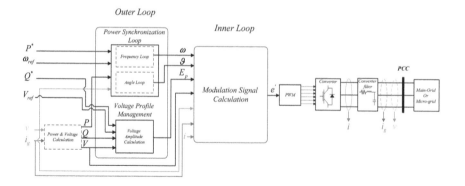

FIGURE 5.4   Generalized control structure of a grid-forming converter [13].

The controller inputs include the converter current, grid current, and voltage at PCC, which are shown as i, i_g, and v, respectively. Also, Other inputs include P^*, Q^*, ω_{ref}, and V_{ref}.

According to this figure, two external and internal control loops are defined. While the outer loop is responsible for calculating the angle, frequency, and voltage amplitude, the inner loop is responsible for determining the modulation signal for the switching blocks of the inverter. In the following, additional explanations and types of methods for the inner and outer layers are provided.

5.4.2   OUTER LOOP: POWER SYNCHRONIZATION LOOP

In the power synchronizer loop, which is shown in Figure 5.4, the goal is to determine the frequency and angle. In the previously published papers, several methods have been proposed for this topic. These methods can be categorized into five groups. The equations of the theta and its control structures are shown in Table 5.1 for each method.

The first method is droop control [13,14]. In this approach, D_f is the droop coefficient, which represents the frequency changes based on the difference between the active reference power (P^*) and the measured power of the converter (P). In the next step, theta is obtained from the integral. This method is simple and does not require the use of additional blocks like phase-locked loop (PLL) for synchronization under normal operating conditions.

The second method is power synchronization control [13,15]. Synchronization is achieved by emulating the power synchronization mechanism of a synchronous machine. In this procedure, the use of PLL blocks is not required under normal grid operation, but the use of PLL blocks is suggested for initial conditions as well as fault conditions.

The third method is enhanced direct power control [13,16]. This method uses PLL permanently. In this method, G, J, and R are gain of the controller, time constant of the controller, and inverse of a frequency droop factor, respectively. Also, J provides a similar behavior of a synchronous machine.

TABLE 5.1

Methods for Implementing Power Synchronization Loop

N	Outer Loop: Power Synchronization Loop	Equation of Theta

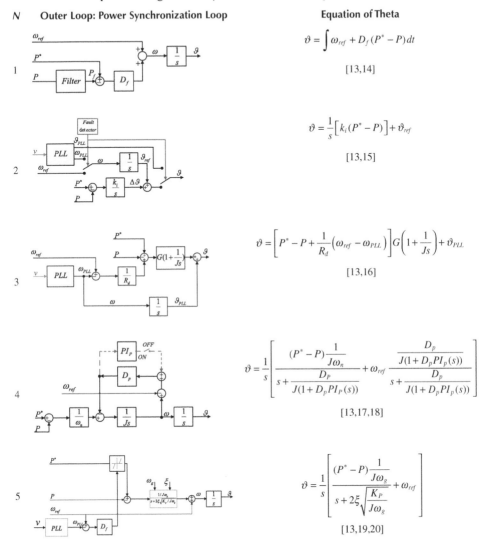

Row 1:
$$\vartheta = \int \omega_{ref} + D_f(P^* - P)dt$$

[13,14]

Row 2:
$$\vartheta = \frac{1}{s}\left[k_i(P^* - P)\right] + \vartheta_{ref}$$

[13,15]

Row 3:
$$\vartheta = \left[P^* - P + \frac{1}{R_d}\left(\omega_{ref} - \omega_{PLL}\right)\right]G\left(1 + \frac{1}{Js}\right) + \vartheta_{PLL}$$

[13,16]

Row 4:
$$\vartheta = \frac{1}{s}\left[\frac{(P^* - P)\frac{1}{J\omega_n}}{s + \frac{D_P}{J(1 + D_p PI_p(s))}} + \omega_{ref}\frac{\frac{D_p}{J(1 + D_p PI_p(s))}}{s + \frac{D_p}{J(1 + D_p PI_p(s))}}\right]$$

[13,17,18]

Row 5:
$$\vartheta = \frac{1}{s}\left[\frac{(P^* - P)\frac{1}{J\omega_g}}{s + 2\xi\sqrt{\frac{K_P}{J\omega_g}}} + \omega_{ref}\right]$$

[13,19,20]

The fourth method is enhanced synchronverter [13,17,18]. It is applicable for both pre-synchronization purposes and normal operation. The constant D_p represents the virtual damping factor of the control and also the steady-state droop. In addition, to improve the dynamic response of the controller, the PI controller is added to the droop.

The fifth method is enhanced synchronous power control [13,19,20]. It includes the second-order transfer function and additional frequency droop loop. In transfer function, J is the virtual moment of inertia, ζ is a damping factor, and K_p represents the steady-state value of the transfer function.

5.4.3 OUTER LOOP: VOLTAGE PROFILE MANAGEMENT

In the voltage profile management loop, which is shown in Figure 5.4, the goal is to determine the voltage amplitude. The voltage profile management can be divided into three groups. The equation of the voltage amplitude and its control structure is shown in Table 5.2 for each method.

TABLE 5.2
Voltage Profile Management Loop

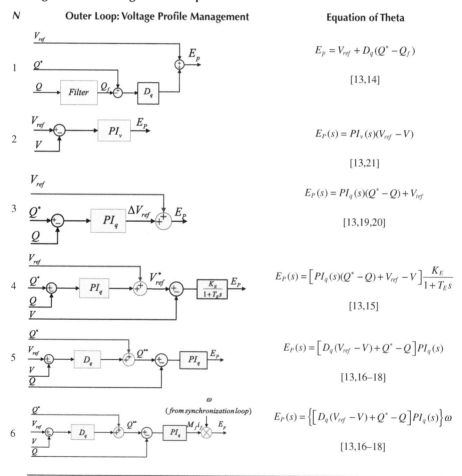

N	Outer Loop: Voltage Profile Management	Equation of Theta
1		$E_p = V_{ref} + D_q(Q^* - Q_f)$ [13,14]
2		$E_P(s) = PI_v(s)(V_{ref} - V)$ [13,21]
3		$E_P(s) = PI_q(s)(Q^* - Q) + V_{ref}$ [13,19,20]
4		$E_P(s) = \left[PI_q(s)(Q^* - Q) + V_{ref} - V\right]\dfrac{K_E}{1 + T_E s}$ [13,15]
5		$E_P(s) = \left[D_q(V_{ref} - V) + Q^* - Q\right]PI_q(s)$ [13,16–18]
6		$E_P(s) = \left\{\left[D_q(V_{ref} - V) + Q^* - Q\right]PI_q(s)\right\}\omega$ [13,16–18]

The first method is droop control [13,14], which is similar to the droop controllers in the synchronization loop. In this equation, D_q is the droop coefficient, which represents the voltage changes based on the difference between the reactive reference power (Q^*) and the measured reactive power of the converter (Q).

The second method is PI-based controller, which is reported in two schemes. One of them is based on the regulation of the voltage [13,21], and the other is based on the regulation of the reactive power with the additional feedforward of the reference voltage [13,19,20].

The third method is cascaded controller [13,15], which is reported in three schemes. This scheme is based on the PI controller and Alternating Voltage controller (AVC).

In addition, the fourth is based on droop controller and PI controller. Finally, the fifth method is similar to the previous one. However, the ω, which is calculated in the synchronization loop, is added to it [13,16–18].

5.4.4 INNER LOOP: CALCULATION OF MODULATION SIGNALS

A wide range of voltage control schemes have been developed for the calculation of modulation signals. The conventional droop-based cascade PI control used for parallel inverters is shown in Figure 5.5. Unlike vast use of the PI controllers in designing inner and outer control loops for tracking voltage and current set-points of the grid-forming converters, they suffer from serious disadvantages in cascaded operations. First of all, the PI controllers require a precise model of the system in order to minimize the harmonic distortions of inverter voltage and current. On the other hand, dynamic response of these controllers is slow in comparison with the other types of voltage and current controllers. Furthermore, stability margin of these controllers in low switching frequencies is restrict [22]. Consequently, these drawbacks resulted in designing other types of cascade controller for grid-forming converters. The authors

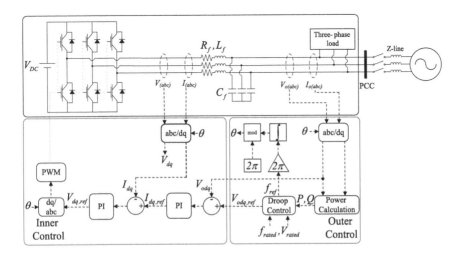

FIGURE 5.5 Conventional droop-based cascaded PI control used for parallel inverters.

of [22] have applied Adaptive Backstepping Integral Nonsingular Fast Terminal Sliding Mode Control (ABINFTSMC) for current controller to follow the reference currents obtained from the voltage controller. Also, the outer loop consisting of voltage controller, which is mixed of H_2/H_∞ type. Furthermore, in [23], the predictive control is presented. Also, neural-network-based methods can be implemented to the grid-forming converters for enhancing their operation [24].

In this chapter, it is tried to introduce a control framework with sliding mode controllers (SMCs). For this purpose, both the inner and the outer control loops are designed via SMC. Also, direct droop control is utilized for specifying set-points of voltage and frequency. So, designing procedure of the SMC for current and voltage controllers is described in the following.

5.4.4.1   Cascade Control by Using Sliding Mode Controller for Current Controller

The cascade droop-based sliding mode control used for parallel inverters is shown in Figure 5.6. In order to design a current controller based on the robust SMC, state-space model of the grid-forming converter is required, which can be derived as equation (5.5) in the dq coordinate:

$$\frac{di_d}{dt} = -\frac{R_f}{L_f}i_d + \omega i_q + V_{DC}\frac{Sw_d}{L_f} - \frac{1}{L_f}V_{od} + \Delta_d \quad \frac{di_q}{dt} = -\frac{R_f}{L_f}i_q - \omega i_d + V_{DC}\frac{Sw_q}{L_f} - \frac{1}{L_f}V_{oq} + \Delta_q$$

$$(5.5)$$

In equation (5.5), $i_{d,q}$ and $V_{od,oq}$ refer to the measured current and output voltage of the converter. In addition, R_f and L_f denote the inverter filters, ω shows the angular frequency which is the output of direct droop. Also, V_{DC} represents the constant DC voltage of the grid-forming inverter. Finally, $Sw_{d,q}$ and $\Delta_{d,q}$ are, respectively, state of the inverter switches and the filter uncertainties in synchronous reference frame.

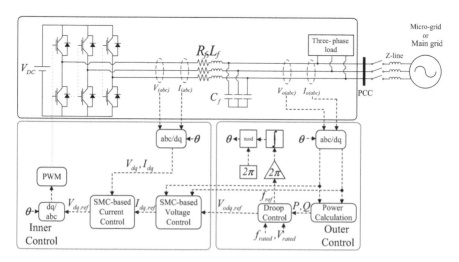

FIGURE 5.6   Cascaded droop-based sliding mode control used for parallel inverters.

In the next step, the sliding surface of the current controller should be determined. In this chapter, the sliding surface in dq frame is considered as equation (5.6):

$$S_d = \left(i_{d,ref} - i_d\right) + C\int_0^t \left(i_{d,ref} - i_d\right)d\tau \quad S_q = \left(i_{q,ref} - i_q\right) + C\int_0^t \left(i_{q,ref} - i_q\right)d\tau \quad (5.6)$$

Then, the input control signals of this controller are considered as below:

$$V_d = Sw_d.V_{DC} \quad V_q = Sw_q.V_{DC} \tag{5.7}$$

Next, the control law consisting of equivalent control and switching function are selected as equations (5.8) and (5.9). These functions draw and keep the system states on the predetermined sliding surfaces by meeting the $S_{d,q} = 0$ and $\dfrac{dS_{d,q}}{dt} = 0$:

$$V_{d,eq} = R.i_d - \omega L i_q + V_{od} + C.L_f\left(i_{d,ref} - i_d\right)$$

$$V_{q,eq} = R.i_q + \omega L i_d + V_{oq} + C.L_f\left(i_{q,ref} - i_q\right)$$

(5.8)

$$V_d^s = K.sign\left(S_d\right) \quad V_q^s = K.sign\left(S_q\right) \tag{5.9}$$

where K refers to the switching gain of the current controller. The modulation signal of the converter in the current controller can be achieved by adding up the corresponding equations of equations (5.8) and (5.9) as equation (5.10). Also, block diagram of the inner control loop is illustrated in Figure 5.7:

$$V_d^{ref} = R_f.i_d - \omega.L_f.i_q + V_{od} + C.L_f\left(i_{d,ref} - i_d\right) + K.sign\left(S_d\right)$$

$$V_q^{ref} = R_f.i_q + \omega.L_f.i_d + V_{oq} + C.L_f\left(i_{q,ref} - i_q\right) + K.sign\left(S_q\right)$$

(5.10)

For designing the sliding-mode-based voltage controller, the state-space equations of the voltage are written as equation (5.11):

$$\frac{dV_{od}}{dt} = \frac{1}{C_f}\left(i_d - i_{od}\right) + \omega.V_{oq} + \delta_d \quad \frac{dV_{oq}}{dt} = \frac{1}{C_f}\left(i_q - i_{oq}\right) - \omega.V_{od} + \delta_q \quad (5.11)$$

In equation (5.11), the C_f represents the capacitor filter of the inverter, and $\delta_{d,q}$ shows the existing uncertainties. The sliding surface for the voltage controller is considered to be as equation (5.12):

$$\sigma_d = \left(V_{od,ref} - V_{od}\right) + C'\int_0^t \left(V_{od,ref} - V_{od}\right)d\tau \quad \sigma_q = \left(V_{oq,ref} - V_{oq}\right) + C'\int_0^t \left(V_{oq,ref} - V_{oq}\right)d\tau$$

(5.12)

Furthermore, the i_d and i_q are taken as the input control signal of this SMC-based voltage controller. On the other hand, by considering $\sigma_{d,q} = 0$ and $\dfrac{d\sigma_{d,q}}{dt} = 0$, the

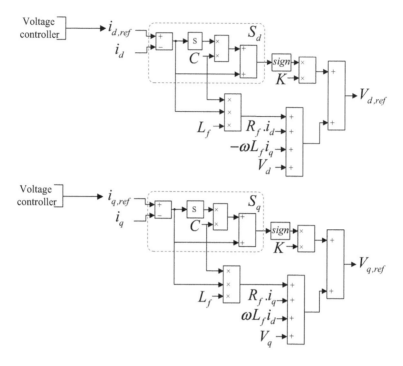

FIGURE 5.7   Control block diagram of the SMC-based current controller.

equivalent controller and switching functions are obtained as equations (5.13) and (5.14):

$$i_{d,eq} = C_f \frac{dV_{od,ref}}{dt} + i_{od} - \omega.C_fV_{oq} + C_f.C'\left(V_{od,ref} - V_{od}\right)$$

$$i_{q,eq} = C_f \frac{dV_{oq,ref}}{dt} + i_{oq} + \omega.C_fV_{od} + C_f.C'\left(V_{oq,ref} - V_{oq}\right)$$

$$\tag{5.13}$$

$$i_d^s = K'.sign\left(\sigma_d\right)\ i_q^s = K'.sign\left(\sigma_q\right) \tag{5.14}$$

Finally, adding the correspondence components of equations (5.13) and (5.14) results in $i_{d,ref}$ and $i_{q,ref}$, which is shown in equation (5.15) and Figure 5.8:

$$i_{d,ref} = C_f \frac{dV_{od,ref}}{dt} + i_{od} - \omega.C_fV_{oq} + C_f.C'\left(V_{od,ref} - V_{od}\right) + K'.sign\left(\sigma_d\right)$$

$$i_{q,ref} = C_f \frac{dV_{oq,ref}}{dt} + i_{oq} + \omega.C_fV_{od} + C_f.C'\left(V_{oq,ref} - V_{oq}\right) + K'.sign\left(\sigma_q\right)$$

$$\tag{5.15}$$

The presented control strategy is evaluated via different scenarios to assess its performance in terms of accuracy and dynamic response. First of all, a heavy step load is

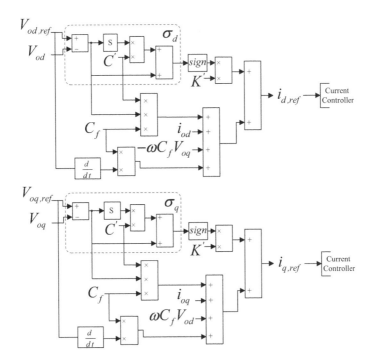

FIGURE 5.8   Control block diagram of the SMC-based voltage controller.

connected to the system at $t=0.9\,\text{s}$. The voltage and current waveforms beside active power output and frequency of the grid-forming inverter are depicted in Figure 5.9.

The next scenario is allocated to the grid-connected mode in which a 30% three-phase voltage drop occurs in the main-grid side at $t=0.9\,\text{s}$, which lasts for $0.3\,\text{s}$. This scenario is important since nowadays the grid-connected power sources should be capable of riding through capabilities during abnormal and severe grid conditions because of newly established requirements for grid codes such as Low-Voltage-Ride-Through (LVRT). Figure 5.10 highlights the performance of the grid-forming inverter during this large signal event.

During this scenario despite occurring a large signal voltage drop, the DG was able to inject reactive power in order to support the main grid, which ended up in compensating the voltage magnitude at the point-of-common-coupling. Consequently, the active power in the inverter side is injected properly as pre-fault situation.

Analyzing the results achieved from implementing the presented controller clears that this controller performs well during small and large signal perturbations in terms of accuracy and dynamic response.

5.4.4.2   Direct Voltage Control by Using Sliding Mode Controller

Typically, in single-source grids with inductance-capacitor (*LC*) or inductance-capacitor-inductive (*LCL*) filters, implementing cascade control using pulse width modulation is more common. However, the slow response of the outer loop voltage

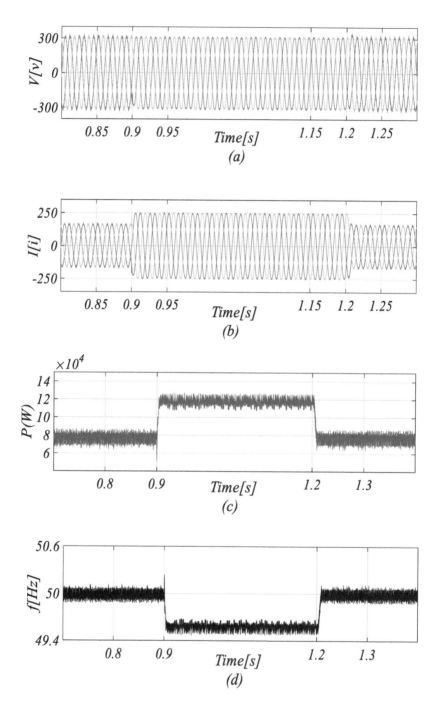

FIGURE 5.9 Performance of the grid-forming inverter during step load change scenario. (a) Voltage waveform. (b) Current waveform. (c) Active power output. (d) Frequency.

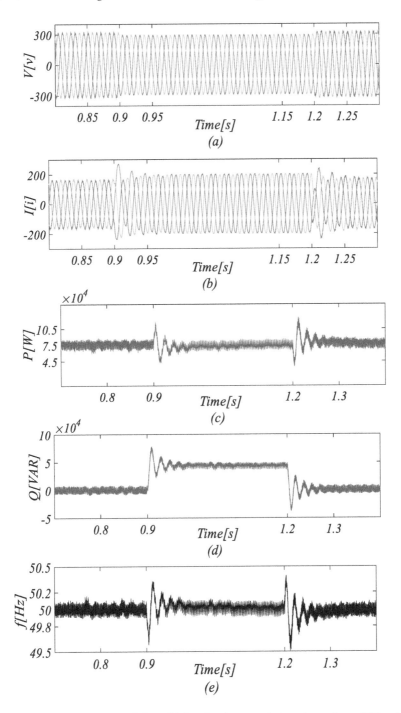

FIGURE 5.10   Performance of the grid-forming inverter during three-phase 30% voltage sag. (a) Voltage waveform. (b) Current waveform. (c) Active power output. (d) Reactive power output. (e) Frequency.

controller is a drawback. So, it is an obstacle for changing the reference current while the load current changes abruptly [25]. Consequently, to fill the gap mentioned above, the direct voltage control using sliding mode concept is introduced as an alternative for cascaded controllers by some of the studies.

For example, authors in [25] have proposed a direct voltage controller using SMC. The presented framework consists of three parts. The first component is a feedforward controller which is designed for ideal situation and steady-state operation of the system. Also, this strategy benefits from a discontinuous SMC for modeling the uncertainties of the system, compensating disturbances, and transient loads. Finally, the feedback control acts as a damper and speeds up the dynamic response.

Also in [26], it is mentioned that the conventional SMC might have a long settling time. So, a fast terminal SMC is presented in this study for active power sharing during transient states. Moreover, unbalanced situations are considered in designing this controller. As a result, a negative sequence part is embedded to the presented controller to tackle the unbalanced loading situations. The comparative results show good performance of this controller, because the overshoot level of frequency and active power of the presented method are considerably lower than the results achieved by linear PI controllers during different scenarios.

Furthermore, authors of [27] have introduced an adaptive method based on SMC for the isolated microgrids operating in Master/Slave mode. The control strategy of the master unit, which operates in voltage control mode, is designed such that it is independent of system topology, parameters, and dynamics of microgrid loads. In addition, this method is robust against load disturbances and sudden disconnection of the slave units. Analysis of the results demonstrates that this method is robust against parametric uncertainties and performs well during unbalanced and harmonic load energization.

Designing procedure of the direct voltage control for the grid-forming converters based on the sliding mode concept is provided in this section and in Figure 5.11. To this end, the state-space equations of the converters in the stationary reference frame can be written as:

$$\frac{di_{\alpha\beta}}{dt} = -\frac{R_f}{L_f}i_{\alpha\beta} - \frac{1}{L_f}V_{o,\alpha\beta} + \frac{1}{L_f}V_{\alpha\beta}$$

$$\frac{dV_{o,\alpha\beta}}{dt} = \frac{1}{C_f}i_{\alpha\beta} - \frac{1}{C_f}i_{o,\alpha\beta}$$

(5.16)

Also, the sliding surface is defined as equation (5.17):

$$S_\alpha = \frac{d}{dt}\left(V_{o,ref,\alpha} - V_{o,\alpha}\right) \quad S_\beta = \frac{d}{dt}\left(V_{o,ref,\beta} - V_{o,\beta}\right)$$

(5.17)

In order to achieve the equivalent controller, the sliding surface derivatives must be zero. Also, the switching function inputs are determined like the ones selected for cascade controllers. Equation (5.18) shows the equivalent control and switching function of direct-voltage-control-based SMC.

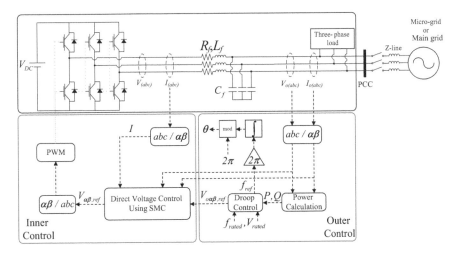

FIGURE 5.11   Inner and outer control block diagram of the direct voltage control using SMC.

$$V_{ref,\alpha\beta}^{eq} = L_f C_f \frac{d^2}{dt^2} V_{o,ref,\alpha\beta} + R_f i_{\alpha\beta} + V_{o,\alpha\beta} + L_f \frac{di_{o,\alpha\beta}}{dt} \quad V_{ref,\alpha\beta}^s = \frac{1}{L_f C_f} K''.sign\left(S_{\alpha\beta}\right)$$

$$(5.18)$$

Consequently, the modulation signals can be obtained as equation (5.19) and shown using Figure 5.12:

$$V_{ref,\alpha} = L_f C_f \frac{d^2}{dt^2} V_{o,ref,\alpha} + \underbrace{\left(R_f i_\alpha + V_{o,\alpha} + L_f \frac{di_{o,\alpha}}{dt} + \frac{1}{L_f C_f} K''.sign\left(S_\alpha\right) \right)}_{-L_f C_f \frac{d^2 V_{o,\alpha}}{dt^2}}$$

$$(5.19)$$

$$V_{ref,\beta} = L_f C_f \frac{d^2}{dt^2} V_{o,ref,\beta} + \underbrace{\left(R_f i_\beta + V_{o,\beta} + L_f \frac{di_{o,\beta}}{dt} + \frac{1}{L_f C_f} K''.sign\left(S_\beta\right) \right)}_{-L_f C_f \frac{d^2 V_{o,\beta}}{dt^2}}$$

The stability analysis of this controller is done via Lyapunov function, which is permanently positive. So, the derivative of the Lyapunov function must be a negative amount to keep the controller in a safe margin. Equation (5.20) shows the mentioned function and its derivative.

$$L = \frac{1}{2} S.S^T > 0 \quad \frac{dL}{dt} < 0 \rightarrow S.S < 0 \qquad (5.20)$$

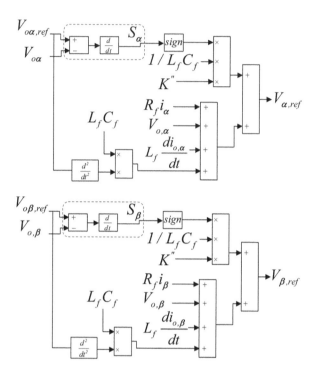

FIGURE 5.12 Inner control block diagram of the direct voltage control using SMC.

So, the following equations can be written, in which $\Delta_{\alpha\beta}$ represents the system uncertainties.

$$
S_{\alpha\beta} \cdot \left(\left[\frac{d^2 V_{o,ref,\alpha\beta}}{dt^2} + \frac{R_f}{L_f C_f} i_{\alpha\beta} + \frac{1}{L_f C_f} V_{o,\alpha\beta} + \frac{1}{C_f} \frac{di_{o,\alpha\beta}}{dt} - \underbrace{\frac{1}{L_f C_f} V_{\alpha\beta}}_{V_{ref,\alpha\beta}} \right] - \Delta_{\alpha\beta} \right) < 0
$$

(5.21)

$$
S_{\alpha\beta} \cdot \left(-K''.sign \left(S_{\alpha\beta} \right) - \Delta_{\alpha\beta} \right) < 0
$$

(5.22)

Consequently, equation (5.23) must be met to have a controller operating in the stable margin.

$$
K'' > \left| \Delta_{\alpha\beta} \right|
$$

(5.23)

The direct voltage control strategy presented in equations (5.16) to (5.19) is tested during two scenarios. First, a 50% mismatch in inductance filter is implemented at $t = 0.5–0.7$ s to evaluate robustness of the controller in terms of parametric

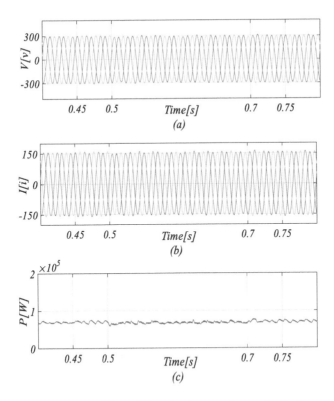

FIGURE 5.13 Performance of the grid-forming inverter during 50% inductance mismatch. (a) Voltage waveform. (b) Current waveform. (c) Active power output.

uncertainties. For this purpose, voltage and current waveforms of the grid-forming converter beside its output active power are highlighted in Figure 5.13.

Analyzing Figure 5.13 proves that the performance of the proposed control method is acceptable since the system maintains its stability.

The control system should be robust facing with nonlinear loads. For this purpose, in the second scenario, a heavy nonlinear RL load ($3.5\Omega + 6\,\mathrm{mH}$) is implemented in the isolated microgrid during $t=0.5$–$0.7\,\mathrm{s}$. Figure 5.14 depicts the performance of the system while a harmonic load connects. Assessing the voltage waveform and magnitude clarifies that the converter does not become instable during harmonic load energization; however, total harmonic distortion (THD) of the output current increases from 1% to 13%, which relatively is a high amount. On the other hand, THD of the voltage does not differ considerably when nonlinear load connects. This amount is under 2%, which is a reasonable value according to IEEE standard 1547.

5.4.4.3 Direct Voltage Control by Using Fractional-Order Sliding Mode Controller

The usage of fractional-order controllers has been expanding in recent years because of their fast response and freedom in selecting controller order. However, to the best of the authors' knowledge, fractional-order controllers based on the sliding mode

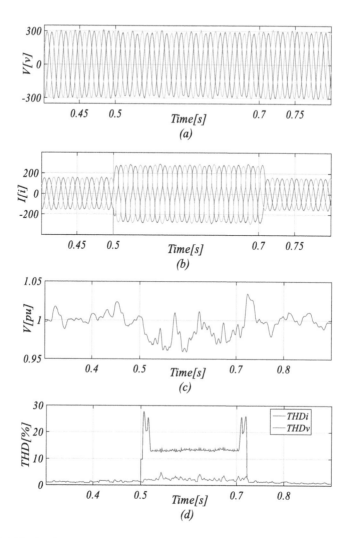

FIGURE 5.14 Performance of the grid-forming with nonlinear load. (a) Voltage waveform. (b) Current waveform. (c) Voltage magnitude of the converter. (d) THD of voltage and current waveforms.

concept have not been studied for grid-forming converters. For this purpose, this subsection aims at designing a robust fractional order SMC (FOSMC) to compare it with the conventional SMC from different points of view. The control block diagram of the direct voltage control using fractional-order sliding mode control is shown in Figure 5.15.

The main difference between the FOSMC and SMC is the fractional order of the sliding surface, which can be a noninteger value in case of FOSMC. So, fractional-order transfer functions are required, which can be shown as $_aD_t^\lambda$ operator [28]. Equation (5.24) describes the mentioned operator clearly.

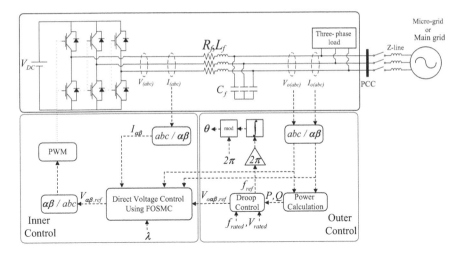

FIGURE 5.15 Inner and outer control block diagram of the direct voltage control using FOSMC.

$$_aD_t^\lambda = \begin{cases} \dfrac{d^\lambda}{dt^\lambda} & \lambda > 0 \\ 1 & \lambda = 0 \\ \displaystyle\int_a^t (d\tau)^\lambda & \lambda < 0 \end{cases}$$ (5.24)

where "a" and "t" show the operation boundaries, and λ is assumed to be the fractional order. There are various ways for estimating the fractional transfer function. For example, in this study, equations (5.25) to (5.26) are utilized [29].

$$G(s) = S^\lambda, \lambda \in R$$ (5.25)

$$\hat{G}(s) = K \prod_{k=-N}^{N} \frac{s + \omega_k'}{s + \omega_k}$$ (5.26)

In equation (5.26), N represents the estimation order. Moreover, ω_k' and ω_k show the zeros and poles of the estimated transfer function:

$$\omega_k' = \omega_b \cdot \left(\frac{\omega_h}{\omega_b}\right)^{\frac{(k+N+0.5(1+\lambda))}{(2N+1)}}$$

$$\omega_k = \omega_b \cdot \left(\frac{\omega_h}{\omega_b}\right)^{\frac{(k+N+0.5(1-\lambda))}{(2N+1)}}$$ (5.27)

$$K = \omega_h^\lambda$$

In equation (5.27), ω_b and ω_h are, respectively, the lower and higher frequencies, which are considered 0.0003 and 5000 Hz. Also, the amount of N is set to be five. For instance, the estimated transfer function of $S^{-0.76}$ is achieved as below:

$$S^{-0.76} = \frac{1.585s^5 + 2350s^4 + 0.0008537s^3 + 0.0007786s^2 + 1783s + 1}{s^5 + 1783s^4 + 0.0007786s^3 + 0.0008537s^2 + 2350s + 1.585}$$

In the following, designing steps of this controller are described. First, the sliding surface is considered as equation (5.28):

$$S_\alpha = \frac{d^\lambda}{dt^\lambda}\left(V_{o,ref,\alpha} - V_{o,\alpha}\right) \quad S_\beta = \frac{d^\lambda}{dt^\lambda}\left(V_{o,ref,\beta} - V_{o,\beta}\right) \tag{5.28}$$

Then, by using the state-space equations of equation (5.16) and meeting the conditions of equation (5.29), the equivalent controller can be obtained as equation (5.30):

$$\frac{dS_{\alpha\beta}}{dt} = 0 \rightarrow \frac{d^{\lambda+1}}{dt^\lambda}\left(V_{o,ref,\alpha\beta}\right) - \frac{d^{\lambda+1}}{dt^\lambda}\left(V_{o,\alpha\beta}\right) = 0 \tag{5.29}$$

$$V_{ref,\alpha}^{eq} = \frac{d^{\lambda+1}}{dt}\left(V_{ref,o,\alpha}\right) + \frac{d^{\lambda-1}}{dt}\underbrace{\left(R_f i_\alpha + V_{o,\alpha} + L_f \frac{di_{o,\alpha}}{dt}\right)}_{\frac{d^2 V_{o,\alpha}}{dt^2}}$$

$$\tag{5.30}$$

$$V_{ref,\beta}^{eq} = \frac{d^{\lambda+1}}{dt}\left(V_{ref,o,\beta}\right) + \frac{d^{\lambda-1}}{dt}\underbrace{\left(R_f i_\beta + V_{o,\beta} + L_f \frac{di_{o,\beta}}{dt}\right)}_{\frac{d^2 V_{o,\beta}}{dt^2}}$$

Also, the switching function of this control is just like the one designed for direct voltage control using SMC in the previous part.

Finally, by adding the switching function, the modulation signal of the grid-forming converter can be written as equation (5.31) and Figure 5.16.

$$V_{ref,\alpha} = L_f C_f \frac{d^{\lambda+1}}{dt}\left(V_{ref,o,\alpha}\right) + \frac{d^{\lambda-1}}{dt}\left(R_f i_\alpha + V_{o,\alpha} + L_f \frac{di_{o,\alpha}}{dt}\right) + \frac{1}{L_f C_f}K''.\mathrm{sign}\left(S_\alpha\right)$$

$$V_{ref,\beta} = L_f C_f \frac{d^{\lambda+1}}{dt}\left(V_{ref,o,\beta}\right) + \frac{d^{\lambda-1}}{dt}\left(R_f i_\beta + V_{o,\beta} + L_f \frac{di_{o,\beta}}{dt}\right) + \frac{1}{L_f C_f}K''.\mathrm{sign}\left(S_\beta\right)$$

$$\tag{5.31}$$

The stability analysis of the FOSMC is performed in the following equations as well as the SMC.

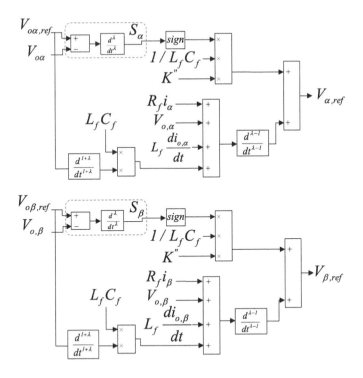

FIGURE 5.16   Inner control block diagram of the direct voltage control using FOSMC.

$$
S_{\alpha\beta} \cdot \left(\left[\begin{array}{c} \dfrac{d^{\lambda+1}V_{o,ref,\alpha\beta}}{dt^2} + \dfrac{d^{\lambda-1}}{dt^{\lambda-1}}\left(\dfrac{R_f}{L_fC_f}i_{\alpha\beta} + \dfrac{1}{L_fC_f}V_{o,\alpha\beta} + \dfrac{1}{C_f}\dfrac{di_{o,\alpha\beta}}{dt} \right) \\[2ex] -\dfrac{1}{L_fC_f}\underbrace{V_{\alpha\beta}}_{V_{ref,\alpha\beta}} \end{array} \right] - \Delta_{\alpha\beta} \right) < 0
$$

$$\tag{5.32}$$

$$
S_{\alpha\beta} \cdot \left(-K'' . \text{sign}\left(S_{\alpha\beta} \right) - \Delta_{\alpha\beta} \right) < 0 \tag{5.33}
$$

$$
K'' > \left| \Delta_{\alpha\beta} \right| \tag{5.34}
$$

In this chapter, the SMC and FOSMSC are used for controlling the grid-forming inverters. By setting the fractional order to an integer value, the FOSMC is changed to the conventional SMC. Therefore, the comparison of the SMC and FOSMC is carried out in the black-start scenario. It is visible that tuning the order of the controller to a proper value can considerably reduce the overshoot and response speed.

A black-start scenario is considered for the performance evaluation of the FOSMC in terms of reaching the steady state. For this purpose, the simulation is repeated with different values of λ to observe the system response for various fractional orders shown in Figure 5.17.

As it is visible, when the control order is set to one (as conventional SMC), the response has severe overshoot with fluctuation. In this case, the response has some error in its steady state. Nevertheless, while tuning the λ to 0.45, the output active power experiences the least amount of overshoot and oscillations. Also, harmonic distortions of the terminal voltage are calculated for different values of λ during normal operation of the system, which is reported in Table 5.3.

It can be concluded from Table 5.3 that the THD percentage reduces when the order of the controller is tuned in the range of 0.45–0.8. Consequently, control system designers have to make a trade-off to select the best fractional order for minimizing the THD and improving the dynamic response of the system. Also, the parameters related to the power and control systems are provided in Table 5.4. Since the load feeding by the grid-forming inverter is resistive, the THD value of output current is similar to the THD of voltage reported in this section.

Chattering phenomenon is the most common issue of the SMC-based controllers, which is caused by nonlinear switching control input like sign function. To solve this issue, several methods are presented in the research papers. For example, tanh function can be used instead of sign function. However, using this approach might cause steady-state error in some cases [30]. Also, higher-order controllers can be

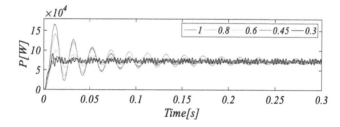

FIGURE 5.17 Active power output of the grid-forming converter using FOSMC for different values of integer and noninteger orders.

TABLE 5.3
THD Values of the System Voltage for Different Values of Fractional Orders

λ	THD v [%]
0.3	4.5
0.45	2.25
0.6	1.3
0.8	0.73
1	1.26

TABLE 5.4

Parameters Used in the Simulations

Parameters of the Power System

V_{DC}	800 V
V_{AC}	380 V
Capacity of the Converter	75 kW
f	50 Hz
f_{sw}	10 kHz
R_f	0.03 Ω
L_f	2 mH
C_f	2,000 μF
Parameters of the Control Systems	
C	650
K	10
C'	50,000
K'	1
K''	1,000

designed to suppress this issue. On the other hand, authors of [31] have declared that the switching gain is proportional to the chattering magnitude. So, in this study, the switching gain is properly selected to reduce the chattering problem.

5.5 SUMMARY

The grid-forming converters are more flexible compared to the grid-following converters. For example, these types of converters can imitate the behavior of synchronous generators; however, this capability is quite impossible in case of grid-following inverters. So, this type of inverter has gained much more attention than their conventional counterparts, in recent years. To this end, in this chapter, different methods of controlling grid-forming converters are discussed. First, it is stated that two control loops are required for the proper operation of these types of converters, which are responsible for synchronizing power (outer loop) and tracking reference values (inner loop). Accordingly, in this chapter, various types of power synchronization methods are reviewed in detail. Second, the methods used for the inner loop are divided into two types: cascade control and direct voltage control. As mentioned in the text, the cascade control is suffering from slower dynamic response. Therefore, direct voltage controllers are much preferable. Moreover, in this chapter, both cascade and direct voltage controller are designed using droop function for the outer loop and SMC for the inner loop of a grid-forming converter. Various scenarios are carried out for performance evaluation including, step load change, voltage sag event, black start, nonlinear load feeding, and filter deviation. The simulation results indicated that the SMC approach combined with droop function has an acceptable operation during the different scenarios. Finally, a FOSMC method is developed for grid-forming converter to compare it with the conventional SMC approaches. It was observable that FOSMC outperforms the SMC method, since setting the fractional order to a proper value reduced the output active power overshoot considerably and enhanced the dynamic response. In addition, the THD value of the converter voltage can be reduced by tuning the fractional order of the controller. The simulations are done via MATLAB/Simulink software.

REFERENCES

[1] T. Ackermann, G. Andersson, and L. Söder, "Distributed generation: A definition," *Electric Power Systems Research*, vol. 57, no. 3, pp. 195–204, 2001.

[2] H. Zhang, W. Xiang, W. Lin, and J. Wen, "Grid forming converters in renewable energy sources dominated power grid: Control strategy, stability, application, and challenges," *Journal of Modern Power Systems and Clean Energy*, vol. 9, no. 6, pp. 1239–1256, 2021.

[3] D. B. Rathnayake et al., "Grid forming inverter modeling, control, and applications," *IEEE Access*, vol. 9, pp. 114781–114807, 2021.

[4] K. Abbaszadeh and S. Roozbehani, "A new approach for maximum power extraction from wind turbine driven by doubly fed induction generator based on sliding mode control," *Energy Engineering & Management*, vol. 1, no. 2, pp. 11–22, 2012.

[5] S. Roozbehani, K. Abbaszadeh, and M. Torabi, *Sensorless Maximum Wind Energy Capture Based on Input Output Linearization and Sliding Mode Control,* in *IET Conference on Renewable Power Generation (RPG 2011)*, Edinburgh, 2011.

[6] N. Bolouki, S. Roozbehani, and K. Abbaszadeh, "Second order sliding mode control of permanent-magnet synchronous wind generator for direct active and reactive power control," in *The 5th Annual International Power Electronics, Drive Systems and Technologies Conference (PEDSTC 2014)*, pp. 434–439, Tehran, Iran,2014.

[7] M. T. Hagh, S. Roozbehani, F. Najaty, S. Ghaemi, Y. Tan, and K. M. Muttaqi, "Direct power control of DFIG based wind turbine based on wind speed estimation and particle swarm optimization," in *2015 Australasian Universities Power Engineering Conference (AUPEC)*, pp. 1–6, Wollongong, NSW, Australia, 2015.

[8] C. Lascu, "Sliding-mode direct-voltage control of voltage-source converters with LC filters for pulsed power loads," *IEEE Transactions on Industrial Electronics*, vol. 68, no. 12, pp. 11642–11650, 2020.

[9] D. Wu, F. Tang, J. C. Vasquez, and J. M. Guerrero, "Control and analysis of droop and reverse droop controllers for distributed generations," *2014 IEEE 11th International Multi-Conference on Systems, Signals and Devices, SSD 2014*, 2014, doi: 10.1109/SSD.2014.6808842.

[10] S. Roozbehani, M. T. Hagh, and S. G. Zadeh, "Frequency control of islanded wind-powered microgrid based on coordinated robust dynamic droop power sharing," *IET Generation, Transmission & Distribution*, vol. 13, no. 21, pp. 4968–4977, 2019.

[11] M. T. Hagh, S. Roozbehani, and S. Ghasemzadeh, "Dynamic reverse droop power sharing in microgrid based on neural networks," *Procedia Computer Science*, vol. 120, pp. 766–779, 2017.

[12] S. Prakash and S. Mishra, "Fast terminal sliding mode control for improved transient state power sharing between parallel VSCs in an autonomous microgrid under different loading conditions," *IET Renewable Power Generation*, vol. 14, no. 6, pp. 1063–1073, 2020.

[13] R. Rosso, X. Wang, M. Liserre, X. Lu, and S. Engelken, "Grid-forming converters: Control approaches, grid-synchronization, and future trends—A review," *IEEE Open Journal of Industry Applications*, vol. 2, pp. 93–109, 2021.

[14] S. D'Arco and J. A. Suul, "Virtual synchronous machines—Classification of implementations and analysis of equivalence to droop controllers for microgrids," in *2013 IEEE Grenoble Conference*, pp. 1–7, Grenoble, France 2013.

[15] L. Zhang, L. Harnefors, and H.-P. Nee, "Power-synchronization control of grid-connected voltage-source converters," *IEEE Transactions on Power systems*, vol. 25, no. 2, pp. 809–820, 2010.

[16] M. Ndreko, S. Rüberg, and W. Winter, "Grid forming control for stable power systems with up to 100% inverter based generation: A paradigm scenario using the IEEE 118-bus system," in *Proceedings of the 17th International Wind Integration Workshop*, Stockholm, Sweden, pp. 16–18, 2018.

[17] Q.-C. Zhong, P.-L. Nguyen, Z. Ma, and W. Sheng, "Self-synchronized synchronverters: Inverters without a dedicated synchronization unit," *IEEE Transactions on Power Electronics*, vol. 29, no. 2, pp. 617–630, 2013.

[18] Q.-C. Zhong and G. Weiss, "Synchronverters: Inverters that mimic synchronous generators," *IEEE Transactions on Industrial Electronics*, vol. 58, no. 4, pp. 1259–1267, 2010.

[19] P. Rodriguez, C. Citro, J. I. Candela, J. Rocabert, and A. Luna, "Flexible grid connection and islanding of SPC-based PV power converters," *IEEE Transactions on Industry Applications*, vol. 54, no. 3, pp. 2690–2702, 2018.

[20] D. Remon, A. M. Cantarellas, E. Rakhshani, I. Candela, and P. Rodriguez, "An active power synchronization control loop for grid-connected converters," in *2014 IEEE PES General Meeting| Conference & Exposition*, pp. 1–5, National Harbor, MD, USA, 2014.

[21] C. Li, R. Burgos, I. Cvetkovic, D. Boroyevich, L. Mili, and P. Rodriguez, "Analysis and design of virtual synchronous machine based STATCOM controller," in *2014 IEEE 15th Workshop on Control and Modeling for Power Electronics (COMPEL)*, pp. 1–6, Santander, Spain, 2014.

[22] M. Raeispour, H. Atrianfar, H. R. Baghaee, and G. B. Gharehpetian, "Robust sliding mode and mixed H-2 H∞ output feedback primary control of AC microgrids," *IEEE Systems Journal*, vol. 15, no. 2, pp. 2420–2431, 2021, doi: 10.1109/JSYST.2020.2999553.

[23] V. Yaramasu, M. Rivera, M. Narimani, B. Wu, and J. Rodriguez, "Model predictive approach for a simple and effective load voltage control of four-leg inverter with an output LC filter," *IEEE Transactions on Industrial Electronics*, vol. 61, no. 10, pp. 5259–5270, 2014, doi: 10.1109/TIE.2013.2297291.

[24] X. Fu and S. Li, "Control of single-phase grid-connected converters with LCL Filters using recurrent neural network and conventional control methods," *IEEE Transactions on Power Electronics*, vol. 31, no. 7, pp. 5354–5364, 2016, doi: 10.1109/TPEL.2015.2490200.

[25] C. Lascu, "Sliding-mode direct-voltage control of voltage-source converters with LC Filters for pulsed power loads," *IEEE Transactions on Industrial Electronics*, vol. 68, no. 12, pp. 11642–11650, 2021, doi: 10.1109/TIE.2020.3040694.

[26] S. Prakash and S. Mishra, "Fast terminal sliding mode control for improved transient state power sharing between parallel VSCs in an autonomous microgrid under different loading conditions," *IET Renewable Power Generation*, vol. 14, no. 6, pp. 1063–1073, 2020, doi: 10.1049/iet-rpg.2019.0621.

[27] M. M. Rezaei and J. Soltani, "Robust control of an islanded multi-bus microgrid based on input-output feedback linearisation and sliding mode control," *IET Generation, Transmission and Distribution*, vol. 9, no. 15, pp. 2447–2454, 2015, doi: 10.1049/iet-gtd.2015.0340.

[28] N. Bouarroudj, D. Boukhetala, and F. Boudjema, "Sliding-mode controller based on fractional order calculus for a class of nonlinear systems," *International Journal of Electrical and Computer Engineering*, vol. 6, no. 5, pp. 2239–2250, 2016, doi: 10.11591/ijece.v6i5.pp2239–2250.

[29] A. Oustaloup, F. Levron, B. Mathieu, and F. M. Nanot, "Frequency-band complex non-integer differentiator: Characterization and synthesis," *IEEE Transactions on Circuits and Systems I: Fundamental Theory and Applications*, vol. 47, no. 1, pp. 25–39, 2000, doi: 10.1109/81.817385.

[30] M. P. Aghababa, "Design of a chatter-free terminal sliding mode controller for nonlinear fractional-order dynamical systems," *International Journal of Control*, vol. 86, no. 10, pp. 1744–1756, 2013, doi: 10.1080/00207179.2013.796068.

[31] H. Lee and V. I. Utkin, "Chattering suppression methods in sliding mode control systems," *Annual Reviews in Control*, vol. 31, no. 2, pp. 179–188, 2007, doi: 10.1016/j.arcontrol.2007.08.001.

6 Small-Signal Modeling and Validation including State-Space and Admittance Models of the Virtual Synchronous Machine

Jingzhe Xu, Weihua Zhou, and Behrooz Bahrani
Monash University

CONTENTS

6.1   INTRODUCTION

In the last decades, inverter-based resources (IBRs) have been increasingly employed
to replace the conventional synchronous generators due to the threats of climate
change, fossil fuel shortage, and environment pollution [1]. However, the synchronous generators displacement and IBRs utilization inevitably bring in instability
issues, since most of the current IBRs in service are under the grid-following control
mode, which may cause the rotational inertia reduction and adverse control interactions [2,3].

Originated from supporting microgrids, the grid-forming inverters (GFMIs) seem
to be one of the prominent solutions, since they provide independent voltage sources
forming the grid rather than dependent current sources as the grid-following inverters (GFLIs) behave [4]. Therefore, the stability and other ancillary services under the
decreasing system strength conditions can be guaranteed. Only the state-space models are derived in references [5–9], whereas only the impedance models are compre-
hensively studied in references [10–12]. However, state-space models and impedance
models, as two representations of the small-signal models, are rarely linked in these
existing literature. To provide a handy small-signal model including the state-space
and impedance models for further applications, e.g., controller design, converter-grid
interaction, eigenvalue analysis, and impedance reshaping, this paper presents a unified state-space and impedance model derivation framework of the virtual synchronous machines (VSMs), one of GFMI technology variants. This paper step-by-step
derives both the state-space and admittance models of the VSM. The correctness of
the established small-signal models are validated by frequency scanning and time-
domain step response. In addition, the controller parameter sensitivities to the admittance and eigenvalues loci are investigated.

The rest of this paper is organized as follows. Section 6.2 briefs the system architecture to be modeled and the component connection method (CCM) being employed.
Section 6.3 derives the state-space model at the component level and exemplifies the
combination of subsystems. Section 6.4 validates the built system model in frequency
domain and time domain. Finally, Section 6.5 gives conclusion.

6.2   INTRODUCTION OF THE STUDIED SYSTEM
AND MODELING METHOD

This section describes the studied system and the employed small-signal modeling
methodology, CCM.

6.2.1 System Description

To illustrate the CCM method initially proposed in reference [13] and validate its feasibility in the complete modeling of a GFMI, a simplified system model is established as Figure 6.1. A three-phase two-level inverter is connected to the grid through an *LCL* filter with a bypass resistor. The inverter's control system comprises outer power control loops, inner voltage and current control loops, and modulation delay. The purposes of this control structure are first tracking the terminal voltage of the capacitor V_{C_f} with the generated reference and second injecting the required power to the grid through the point of common coupling (PCC). The voltage reference is either constant or determined by the reactive power controller.

Figure 6.2 presents the overview of the VSM's control structure. It details inner current and voltage controls, outer active and reactive power controls, and reference frame rotations. Blocks in green shades are fundamental components to be modeled in state-spaces accordingly. Other parts are to be established through connection

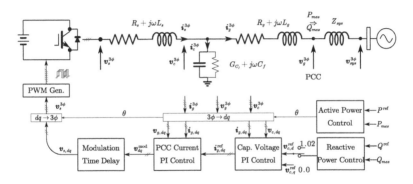

FIGURE 6.1 Single-line diagram of the VSM system and control overview.

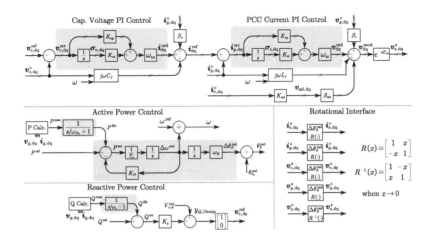

FIGURE 6.2 Overview of VSM control loops.

algorithms. The detailed small-signal model derivation of each subsystem will be described in Section 6.3. One of the highlights is that the model includes the impacts of Park-Clarke-Transformation (PCT) [14] and its reversal at all interfaces as shown in the right bottom of Figure 6.2.

6.3 CCM DESCRIPTION

The CCM combines linear subsystems defined in state-space model into a whole system. The method first stakes state, input, and output variables in row vectors and coefficient matrices on diagonals, such that

$$x_{stac} = \left[x_1 x_2 ... x_n\right]^T, u_{stac} = \left[u_1 u_2 ... u_n\right]^T, y_{stac} = \left[y_1 y_2 ... y_n\right]^T,$$

$$A_{stac} = \text{diag}\left[A_1 A_2 ... A_n\right], B_{stac} = \text{diag}\left[B_1 B_2 ... B_n\right], \tag{6.1}$$

$$C_{stac} = \text{diag}\left[C_1 C_2 ... C_n\right], D_{stac} = \text{diag}\left[D_1 D_2 ... D_n\right]$$

It then defines the inputs and outputs of the combined system and finds algebraic relations mapping staked inputs and outputs to the newly defined ones in matrices, i.e., the stacked inputs and the new outputs are the linear combination of stacked outputs and newly defined inputs as

$$u_{stac} = R_1 y_{stac} + R_2 u_{sys},$$
$$y_{sys} = R_3 y_{stac} + R_4 u_{sys}. \tag{6.2}$$

Consequently, it is possible to find the new coefficient matrices by the matrix algebra at the end as

$$A_{sys} = A_{stac} + B_{stac} R_1 \left(I - D_{stac} R_1\right)^{-1} C_{stac},$$

$$B_{sys} = B_{stac} R_1 \left(I - D_{stac} R_1\right)^{-1} D_{stac} R_2 + B_{stac} R_2,$$

$$C_{sys} = R_3 \left(I - D_{stac} R_1\right)^{-1} C_{stac}, \tag{6.3}$$

$$D_{sys} = R_3 \left(I - D_{stac} R_1\right)^{-1} D_{stac} R_2 + R_4$$

Figure 6.3 gives the sketch view of components utilized in CCM.

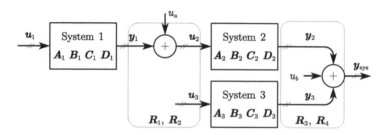

FIGURE 6.3   An overview of CCM methodology.

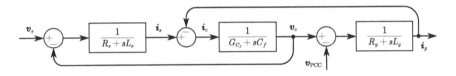

FIGURE 6.4   Single-phase block diagram for grid-connected *LCL* filter.

6.4   SMALL-SIGNAL MODELING OF VSM

6.4.1   *LCL* FILTER MODEL

6.4.1.1   Basic Electrical Model

Figure 6.4 presents the block diagram of a grid-connected *LCL*-filter circuit. Its three-phase equations could be expressed as

$$v_{s,3\phi} - v_{c,3\phi} = \frac{L_s}{\omega_B}\frac{d}{dt}i_{s,3\phi} + R_s i_{s,3\phi},$$

$$v_{c,3\phi} - v_{g,3\phi} = \frac{L_g}{\omega_B}\frac{d}{dt}i_{g,3\phi} + R_g i_{s,3\phi}, \qquad (6.4)$$

$$i_{s,3\phi} - i_{g,3\phi} = \frac{C_f}{\omega_B}\frac{d}{dt}v_{c,3\phi} + G_{C_f} v_{c,3\phi}$$

Further, by transforming the subsystem into $dq-$ frame through PCT under balanced three-phase quantity assumption, the system could be expressed as

$$\frac{L_s}{\omega_B}\frac{d}{dt}i_{s,dq} = -\left(R_s + j\omega L_s\right)i_{s,dq} + \left(v_{s,dq} - v_{c,dq}\right),$$

$$\frac{L_g}{\omega_B}\frac{d}{dt}i_{g,dq} = -\left(R_g + j\omega L_g\right)i_{s,dq} + \left(v_{c,dq} - v_{g,dq}\right), \qquad (6.5)$$

$$\frac{C_f}{\omega_B}\frac{d}{dt}v_{c,dq} = -\left(G_{C_f} + j\omega C_f\right)v_{c,dq} + \left(i_{s,dq} - i_{g,dq}\right),$$

where the state, input, and output variables are defined as $x_{LCL} = \left[i_{s,dq}, i_{g,dq}, v_{c,dq}\right]^T$, $u_{LCL} = \left[v_{s,dq}, v_{g,dq}\right]^T$, and $y_{LCL} = \left[i_{g,dq}, i_{s,dq}, v_{c,dq}\right]^T$, respectively.

6.4.1.2   Resonance Frequency Damping

An ideal *LCL* filter has zero conductance, i.e., $G_{C_f} = 0$. A second-order *LCL* filter has a resonance peak at frequency

$$f_{res} = \frac{1}{2\pi}\sqrt{\frac{L_s + L_g}{L_s L_g C_f}} = \frac{\omega_B}{2\pi}\sqrt{\frac{X_s + X_g}{X_s X_g B_{C_f}}}. \qquad (6.6)$$

One of the passive damping strategies is to parallel a resistor with the filtering capacitor, e.g., $G_{C_f} = 0.01$. The resistor attenuates the resonance peak without altering the lower and higher frequency characters of the filter. However, since the paralleled resistor leads to significant active power consumption, this proposal is not practical. Instead, the capacitor current feedback is one of the common solutions. It replicates the behavior of the resistor by dampening the regulated voltage before PWM processing. The feedback loop is essentially a simple proportional gain, $K_{ad} = G_{C_f} L_s / (K_{PWM} C_f)$, realized by block diagram algebra with pole-zero-cancellation at origin, where $K_{PWM} = 1$ in the per-unitized control loop. The algebraic term for voltage damping is

$$v_{ad,dq} = -K_{ad}\left(i_{s,dq} - i_{g,dq}\right). \tag{6.7}$$

6.4.2 Combined dq – Frame Voltage, Current PI Control, and PWM Delay

6.4.2.1 Current Control Loop

A proportional-integral (PI) controller has the generic transfer function $H(s) = \omega_c\left(sK_p + K_i\right)/s$ with its state-space model expressed as

$$\dot{x} = Ax + Bu,$$

$$y = Cx + Du,$$

where $A = 0$, $B = 1$, $C = \omega_c K_i$, $D = \omega_c K_p$. In addition, K_p, K_i, and ω_c are the proportional gain, integral gain, and closed-loop bandwidth, respectively. Within transformed rotating $dq-$ frame, the AC signals are converted to the DC signals. Therefore, PI controllers with zero tracking error capability may be the right choices. An effective way designing the inner current control loop is to apply zero-placement method. The current control PI zero $s = -K_i/K_p$ approximately cancels the pole introduced by the reactance dynamics, where in this case is the series reactance of the filter till PCC. It is then able to determine the parameters of current control PIs, where $K_{ip} = \omega_B L_\Sigma, K_{ii} = R_\Sigma$, where the loop bandwidth ω_{ic} choice depends on the requirement of converter's current response performance. To reverse the dynamics described by the circuit equation, the controllers' outputs need coupling terms, $j\omega L_s$, and feedforward terms, $\beta_i v_{c,dq}$, where β_i is feedforward coefficient between zero to one for further stability and flexibility. It can then define state variables, inputs, and outputs of the state-space model for the current PI controllers as $x_i = \sigma_{i,dq}, u_i = i_{g,dq}^{err}, y_i = v_{dq}^{ord}$. Coefficient matrices are defined as $A_i = O_{2\times2}, B_i = I_{2\times2}, C_i = K_{ii}I_{2\times2}, D_i = K_{ip}I_{2\times2}$. Figure 6.5 presents the block diagram of the current PI controller considering outer connection relationship.

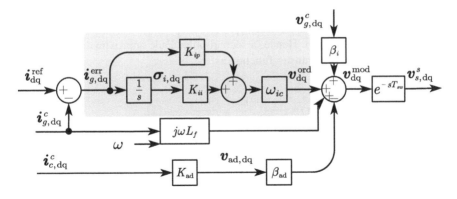

FIGURE 6.5 Block diagram of current PI controller considering outer connection relationship.

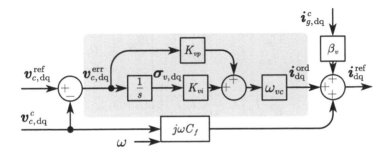

FIGURE 6.6 Block diagram of voltage PI controller considering outer connection relationship.

6.4.2.2 Voltage Control Loop

One of voltage control strategies is regulating the terminal voltage of the capacitor. Similarly, the control strategy aims to reverse the capacitor's dynamics by PI controllers. Reference [15] describes the *Symmetry Optimum* method to tune the parameters. The main idea is to segregate the bandwidth of two controllers by slowing down the voltage PI at least 1/10 of the current PI. The state-space model replicates the structure of PI controllers that state, input, and output variables are as $x_v = \sigma_{v,dq}, u_v = v_{g,dq}^{err}, y_v = i_{dq}^{ord}$, respectively. The coefficient matrices are defined as $A_v = O_{2\times2}, B_v = I_{2\times2}, C_v = K_{vi}I_{2\times2}, D_v = K_{vp}I_{2\times2}$. In terms of fully decomposition, coupling items such as $j\omega C_f$ and $\beta_c i_{g,dq}$ will be taken in account at the output of the controller. Figure 6.6 presents the block diagram of the capacitor voltage PI controller considering outer connection relationship.

6.4.2.3 Padé Approximation

Digital signal processing is the practical implementation for controllers. It inevitably introduces time delay for sampling and processing. The delay time is recognized as 1.5 times of the switching period, T_{sw}, where the A/D conversion introduces a one-step delay and the calculation another half-step delay. The delay $e^{js(1.5T_{sw})}$ is a non-linear transfer function. Hence, a second-order Padé approximation applies to linearize the delay by the transfer function as

$$y(s) = \frac{-0.5sT_{sw} + 1}{0.375(sT_{sw})^2 + sT_{sw} + 1} u(s).$$ (6.9)

Since the switching frequency is high enough for a two-level converter, 10 kHz in this case, the approximation may give proper accuracy for modeling. The state, input, and output variables are defined as $x_{dly} = \begin{bmatrix} \sigma_{d1,d}, \sigma_{d1,q}, \sigma_{d2,d}, \sigma_{d2,q} \end{bmatrix}^T$, $u_{dly} = v_{s,dq}^{mod}$, $y_{dly} = v_{s,dq}$, respectively. The coefficient matrices are

$$A_{dly} = \begin{bmatrix} -\dfrac{8}{3T_{sw}} I_{2\times2} & -\dfrac{8}{3T_{sw}^2} I_{2\times2} \\ I_{2\times2} & O_{2\times2} \end{bmatrix}, B_{dly} = \begin{bmatrix} I_{2\times2} \\ O_{2\times2} \end{bmatrix},$$

$$C_{dly} = \begin{bmatrix} -\dfrac{4}{3T_{sw}} I_{2\times2} & \dfrac{8}{3T_{sw}^2} I_{2\times2} \end{bmatrix}, D_{dly} = O_{2\times2},$$ (6.10)

where **I** is an identity matrix, O is a zero matrix, and subscripts 2×2 is the sizing of the matrix. In later notations, subscripts for 2 by 2 matrices are omitted. At present, all derived subsystems are linear, i.e., coefficient matrices are independent from set-points.

6.4.2.4 Combined Voltage and Current Control Loop

Referring to Figure 6.7, the voltage PI, current PI, and delay approximation connect in series with coupling items and feedforward inputs. If the newly defined control system inputs and outputs are as $u_{vi} = \begin{bmatrix} v_{c,dq}^{ref}, v_{g,dq}, v_{c,dq}, i_{g,dq}, i_{c,dq}, \omega \end{bmatrix}^T$ and $y_{vi} = v_{s,dq}$, respectively, the algebraic relations between the new system and its subsystems are as

$$u_v = v_{c,dq}^{ref} - v_{c,dq},$$

$$u_i = i_{g,dq}^{ref} - i_{g,dq}$$

$$= y_v + j\omega C_f v_{c,dq} + (\beta_v - 1) i_{g,dq},$$

$$u_{dly} = y_i + j\omega L_f i_{g,dq} + \beta_i v_{g,dq} - \beta_{ad} v_{ad,dq}$$ (6.11)

$$= y_i + j\omega L_f i_{g,dq} + \beta_i v_{g,dq} - \beta_{ad} K_{ad},$$

$$y_{vi} = y_{dly}.$$

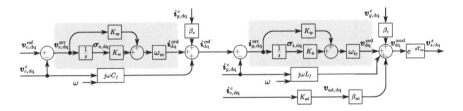

FIGURE 6.7 Block diagram of combined dq-frame voltage, current PI controls, and PWM delay.

Consequently, the linearized model for these connections is as

$$\Delta u_v = \Delta v_{c,dq}^{ref} - \Delta v_{c,dq},$$

$$\Delta u_i = \Delta y_v + jC_f v_{c0,dq}\Delta\omega + j\omega_0 C_f \Delta v_{c,dq} + (\beta_v - 1)\Delta i_{g,dq},$$

$$\Delta u_{dly} = \Delta y_i + jL_f i_{g0,dq}\Delta\omega + j\omega_0 L_f \Delta i_{g,dq} + \beta_i \Delta v_{g,dq} - \beta_{ad}K_{ad}\Delta i_{c,dq},$$

$$\Delta y_{vi} = \Delta y_{dly}.$$

(6.12)

Then it is possible to obtain the relational matrices R_{11} to R_{14} as

$$R_{11} = \begin{bmatrix} O_{2\times4} & O \\ I_{4\times4} & O_{4\times2} \end{bmatrix},$$

$$R_{12} = \begin{bmatrix} I & O & -I & O & O & O_{1\times2} \\ O & O & B_{C_f}J & R_{12}^{3,7}I & O & C_f J v_{c,dq0} \\ O & \beta_i I & O & X_f J & R_{12}^{5,9}I & L_f J i_{g,dq0} \end{bmatrix},$$

(6.13)

$$R_{13} = \begin{bmatrix} O_{2\times4} & I \end{bmatrix}, R_{14} = O_{2\times11},$$

where

$$B_{C_f} = \omega_0 C_f,$$

$$X_f = \omega_0 L_f,$$

$$R_{12}^{3,7} = \beta_v - 1,$$

$$R_{12}^{5,9} = -\beta_{ad}K_{ad},$$

$$J = \begin{bmatrix} 0 & -1 \\ 1 & 0 \end{bmatrix}.$$

With the algebra defined in equations (6.1), (6.2), and (6.3), the new system coefficient matrices are

$$
A_{vi} = \begin{bmatrix}
O & O & O & O \\
k_{vi}I & O & O & O \\
k_{ip}k_{vi}I & k_{ii}I & A_{vi}^{5,5}I & A_{vi}^{5,6}I \\
O & O & I & O
\end{bmatrix},
$$

$$
B_{vi} = \begin{bmatrix}
I & O & -I & O & O & O_{2\times1} \\
k_{vp}I & O & -k_{vp}I + B_{C_f}J & B_{vi}^{3,7}I & O & Jv_{c,dq0} \\
k_{ip}k_{vp}I & \beta_i I & B_{vi}^{5,5}I + B_{vi}^{6,5}J & B_{vi}^{5,7}I + L_f J & R_{12}^{5,9}I & JB_{vi}^{5:6,11} \\
O & O & O & O & O & O_{1\times2}
\end{bmatrix}, \quad (6.14)
$$

$$
C_{vi} = \begin{bmatrix} O_{2\times4} & A_{vi}^{5,5}/2I & -A_{vi}^{5,6}I \end{bmatrix}, D_{vi} = O_{2\times11},
$$

where

$$
k_{vi} = K_{vi}\omega_{vc},
$$

$$
k_{ii} = K_{ii}\omega_{ic},
$$

$$
k_{vp} = K_{vp}\omega_{vc},
$$

$$
k_{ip} = K_{ip}\omega_{ic},
$$

and

$$
A_{vi}^{5,5} = -\frac{8}{3T_{sw}},
$$

$$
A_{vi}^{5,6} = A_{vi}^{5,5}/T_{sw},
$$

$$
B_{vi}^{5,5} = -k_{ip}k_{vp},
$$

$$
B_{vi}^{6,5} = B_{C_f}k_{ip},
$$

$$
B_{vi}^{3,7} = R_{12}^{3,7} = (\beta_v - 1),
$$

$$
B_{vi}^{5,7} = k_{ip}B_{vi}^{3,7},
$$

$$
B_{vi}^{5,9} = R_{12}^{5,9} = -\beta_{ad}K_{ad},
$$

$$
B_{vi}^{5:6,11} = L_f i_{g,dq0} + k_{ip}C_f v_{c,dq0}
$$

6.4.3 SHAFT SWINGING EMULATION

6.4.3.1 Swing Equation

The swing equation applies a variance of the ones for the synchronous generator according to equation (6.5) [14], where T_a mimics two times of inertia constant H. Figure 6.8 presents the block diagram. It provides the synchronous frame for VSM. The generated electrical angle and angular speed are used for PCT and inverse PCT on the voltage and current. Therefore, the perturbation of angle deviation affects the energy exchange on the interface, namely, voltages and currents. The shaft swing can be modeled as

$$\frac{d\Delta\omega^{ord}}{dt} = \frac{1}{T_a}\left(P^{ref} - P_e - K_D\Delta\omega^{ord}\right),$$

$$\frac{d\Delta\theta}{dt} = \omega_B\Delta\omega, \tag{6.15}$$

$$\omega = \omega^{ref} + \Delta\omega,$$

$$\theta = \theta_{S_0} + \Delta\theta.$$

It is then possible to define the state-space model with state, input, and output variables as $x_{SM} = [\omega,\theta]^T$, $u_{SM} = \left[P^{ref},\omega^{ref},P_e\right]^T$, and $y_{SM} = [\omega,\theta]^T$, respectively. Then the correspondent coefficient matrices are

$$A_{SM} = \begin{bmatrix} -\dfrac{K_D}{T_a} & 0 \\ \omega_B & 0 \end{bmatrix},$$

$$B_{SM} = \begin{bmatrix} \dfrac{1}{T_a} & \dfrac{K_D}{T_a} & -\dfrac{1}{T_a} \\ 0 & 0 & 0 \end{bmatrix}, \tag{6.16}$$

$$C_{SM} = I, D = O_{2\times3}.$$

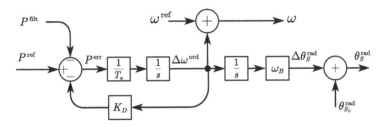

FIGURE 6.8 Block diagram of the swing equation.

6.4.3.2 Power Regulation

Power regulating point is at PCC. In magnitude invariant per-unit $dq-$ frame of VSM, the measured electrical power is calculated as

$$P^{\text{mes}} = 1.5\left(v_{g,d}i_{g,d} + v_{g,q}i_{g,q}\right),$$

$$Q^{\text{mes}} = 1.5\left(v_{g,q}i_{g,d} - v_{g,d}i_{g,q}\right),$$

(6.17)

under a symmetrical three-phase system. The small-signal models are

$$\Delta P^{\text{mes}} = 1.5\left(i_{g,d0}\Delta v_{g,d} + i_{g,q0}\Delta v_{g,q} + v_{g,d0}\Delta i_{g,d} + v_{g,q0}\Delta i_{g,q}\right),$$

$$\Delta Q^{\text{mes}} = 1.5\left(-i_{g,q0}\Delta v_{g,d} + i_{g,d0}\Delta v_{g,q} + v_{g,q0}\Delta i_{g,d} - v_{g,d0}\Delta i_{g,q}\right).$$

(6.18)

Although power calculation is independent from rotating frames, it needs to unify quantities in the same frame. The rotating factors will be processed in the later system assembling. Without a power filter, $P_e = P^{\text{mes}}$.

6.4.3.3 Low-Pass Filter (LPF)

A first-order low-pass filter has a transfer function $y(s) = \dfrac{1}{s/\omega_c + 1}$ as shown in Figure 6.9. Defining the state, input, and output variables as x_{LPF}, u_{LPF}, and y_{LPF}, respectively, the coefficient matrices are $A_{LPF} = -\omega_c$, $B_{LPF} = \omega_c$, $C_{LPF} = 1$, and $D_{LPF} = 0$. The purpose of setting a power filter is to keep the stability of the internal angular speed. It, however, slows down the power tracking. In reference [16], the bandwidth of power filters has been reduced to $5\,\text{Hz}$, i.e., $\omega_{pc} = 10\pi\,\text{rad/s}$.

6.4.3.4 Swing Equation with Power Filter

It may be preferable to connect these two parts as an independent subsystem to the whole control system with the inclusion of power measurement as per equation (6.18) (Figure 6.10). Defining the state variables as $x_P = \left[x_{SM}; x_{P,LPF}\right]$, inputs and outputs as $u_P = [P^{ref}, \omega^{ref}, v_{g,d}, v_{g,q}, i_{g,d}, i_{g,q}]^T$ and $y_P = y_{SM} = [\omega, \theta]^T$, respectively, the connection matrices are

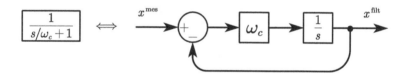

FIGURE 6.9   Block diagram of generic first-order low-pass filter.

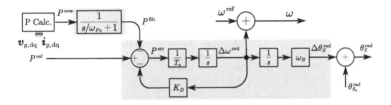

FIGURE 6.10   Block diagram of active power controller with low-pass filter.

$$
R_{21} = \begin{bmatrix} O_{2\times3} \\ O_{1\times2} \quad 1 \\ O_{1\times3} \end{bmatrix},
$$

$$
R_{22} = \begin{bmatrix} I & & & O_{3\times5} \\ O & i_{g,d0} & i_{g,q0} & v_{g,d0} & v_{g,q0} \end{bmatrix}, \tag{6.19}
$$

$$
R_{23} = \begin{bmatrix} I & O_{2\times1} \end{bmatrix}, R_{24} = O_{2\times6}.
$$

The resultant coefficient matrices are

$$
A_P = \begin{bmatrix} -\dfrac{K_D}{T_a} & 0 & 0 \\ \omega_B & 0 & 0 \\ 0 & 0 & -\omega_{Pc} \end{bmatrix},
$$

$$
B_P = \begin{bmatrix} \dfrac{1}{T_a} & \dfrac{K_D}{T_a} & & O_{2\times5} \\ O & i_{g,d0}\omega_{Pc} & i_{g,q0}\omega_{Pc} & \omega_{Pc}v_{g0,R} & \omega_{Pc}v_{g0,I} \end{bmatrix}, \tag{6.20}
$$

$$
C_P = \begin{bmatrix} I & O_{2\times1} \end{bmatrix}, D_P = O_{2\times7}.
$$

6.5   REACTIVE POWER CONTROL

6.5.1   IMPLEMENTATION USING THE LPF AND DROOP CONTROL

Figure 6.11 presents a type of droop control of reactive power control. The connectors are then derived as

$$
\Delta u_{Q,LPF} = \Delta Q_g,
$$

$$
\Delta y_{Q,Droop} = -K_q \Delta y_{Q,LPF} + K_q \Delta Q^{ref} + \Delta V_{c,d}^{ord}. \tag{6.21}
$$

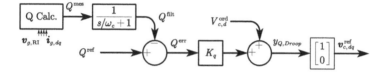

FIGURE 6.11 Reactive power droop control with low-pass filter.

The SSM of this module is defined as state variable $x_Q = x$, inputs $u_Q = \left[v_{g,R}, v_{g,I}, i_{g,d}, i_{g,q}, Q^{ref}, V_{c,d}^{ref} \right]^T$ and output $y_Q = v_{c,d}^{ref}$. The connection matrices are consequently written as

$$R_{31} = 0,$$

$$R_{32} = \begin{bmatrix} -i_{g,q0} & i_{g,d0} & v_{g,q0} & -v_{g,d0} & 0 & 0 \end{bmatrix}, \qquad (6.22)$$

$$R_{33} = -K_q, R_{34} = \begin{bmatrix} 0_{1\times6} & K_q & 1 \end{bmatrix}.$$

Thus, the coefficient matrices are $A_Q = -\omega_{Qc}$, $B_Q = \omega_{Qc} R_2$, $C_Q = -K_q$, and $D_Q = O_{2\times6}$.

6.6 FRAME ROTATION

Recalling the concept of phasor, if a vector x is defined in a rotating Cartesian coordinate with the angular frequency of ω, it can be expressed in polar form as $\vec{x} = |x| e^{j\delta}$. The angular speed disturbance causes the perturbation of the angle. As shown in Figure 6.12, the deviation $\Delta\delta$ varies its orthogonal decomposition on the dq axis or $\vec{x}_{\Delta t} = \vec{x}_0 e^{j\Delta\delta}$. Expanding the equation in matrix form, it could have $\vec{x}_{dq,\Delta t} = R(\Delta\delta)\vec{x}_{dq,0}$, where the rotation function

$$R(\Delta\delta) = \begin{bmatrix} \cos(\Delta\delta) & \sin(\Delta\delta) \\ -\sin(\Delta\delta) & \cos(\Delta\delta) \end{bmatrix}. \qquad (6.23)$$

Furthermore, the reverse of the rotation function is $R^{-1}(\Delta\delta) = \begin{bmatrix} \cos(\Delta\delta) & -\sin(\Delta\delta) \\ \sin(\Delta\delta) & \cos(\Delta\delta) \end{bmatrix} = R^T(\Delta\delta)$. If $\Delta\delta \to 0$, then $\cos(\Delta\delta) \to 1$, and $\sin(\Delta\delta) \to \Delta\delta$. The rotation function and its inverse are approximately simplified to $R(\Delta\delta) \cong \begin{bmatrix} 1 & \Delta\delta \\ -\Delta\delta & 1 \end{bmatrix}$ and $R^{-1}(\Delta\delta) \cong \begin{bmatrix} 1 & -\Delta\delta \\ \Delta\delta & 1 \end{bmatrix}$, respectively.

The electrical angle $\theta(t)$ generated from the swing equation (6.15) is the internal dq reference. When defining the power flow from VSM to PCC positive, the internal

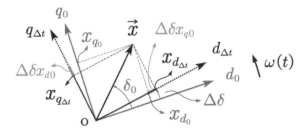

FIGURE 6.12 Illustration of rotating frame disturbances.

frame leads the external voltage reference at PCC from the well-known a.c. power transfer function, $P = V_s V_R \sin\theta / X$. The controlled capacitor voltage is arbitrarily aligned on the d − axis, that is, $v_{c,q} = 0$. The grid voltage phasor is then lagging the VSM's internal dq − frame by the angle of θ. Since all signals processed in the controller are based on the VSM's frame, in general, the captured signals from physical circuits take into account the effects of the rotating frame, e.g., $\mathbf{i}_{g,dq}, \mathbf{v}_{c,dq}$, and $\mathbf{v}_{g,dq}$. On the contrary, signals sent to the real world, e.g., $\mathbf{v}_{s,dq}$, may take effects of the inverse rotation into account.

Using superscript 'c' for signals inside VSM control loops and 's' for signals imposed on the electrical circuit, the SSM for the converter sampling interfaces is

$$\Delta x_d^c = \Delta x_d^s + x_{q0}\Delta\delta,$$
$$\Delta x_q^c = \Delta x_q^s - x_{d0}\Delta\delta. \tag{6.24}$$

Similarly, the SSM for converter electrical interfacing is

$$\Delta x_d^s = \Delta x_d^c - x_{q0}\Delta\delta,$$
$$\Delta x_q^s = \Delta x_q^c + x_{d0}\Delta\delta. \tag{6.25}$$

6.7 DERIVATION AND VALIDATION OF THE WHOLE-SYSTEM SMALL-SIGNAL MODEL

6.7.1 DERIVATION AND VALIDATION OF THE STATE-SPACE MODEL

At this stage, the overall system will be structured from interconnecting established SSM modules, namely, the *LCL* filter, the voltage and current PI controllers, the active and reactive power controllers, and the rotation interface. The state variables, inputs, and outputs of VSM are defined as $x_{VSM} = [x_{LCL}; x_{vi}; x_P; x_Q]$, $u_{VSM} = \left[v_{g,dq}^T, \left(v_{c,dq}^{ref}\right)^T, \omega^{ref}, P^{ref}, Q^{ref}\right]^T$, and $y_{VSM} = i_{g,dq}$, respectively. The establishment of connection matrices is straightforward by mapping interfaces between control modules and the filter. Taking into account the effects brought by angle disturbances as previously determined, interface equations are formed as

$$\Delta i_{g,d}^c = \Delta i_{g,d}^s + i_{g,q0}\Delta\theta,$$

$$\Delta i_{g,q}^c = \Delta i_{g,q}^s - i_{g,d0}\Delta\theta; \tag{6.26}$$

$$\Delta i_{c,d}^c = \Delta i_{c,d}^s + i_{c,q0}\Delta\theta$$

$$= \left(\Delta i_{s,d}^s - \Delta i_{g,d}^s\right) + \left(i_{s,q0} - i_{g,q0}\right)\Delta\theta,$$

$$\Delta i_{c,q}^c = \Delta i_{c,q}^s - i_{c,d0}\Delta\theta$$

$$= \left(\Delta i_{s,q}^s - \Delta i_{g,q}^s\right) - \left(i_{s,d0} - i_{g,d0}\right)\Delta\theta; \tag{6.27}$$

$$\Delta v_{g,d}^c = \Delta v_{g,d}^s + v_{g,q0}\Delta\theta,$$

$$\Delta v_{g,q}^c = \Delta v_{g,q}^s - v_{g,d0}\Delta\theta; \tag{6.28}$$

$$\Delta v_{c,d}^c = \Delta v_{c,d}^s + v_{c,q0}\Delta\theta,$$

$$\Delta v_{c,q}^c = \Delta v_{c,q}^s - v_{c,d0}\Delta\theta; \text{ and,} \tag{6.29}$$

$$\Delta v_{s,d}^s = \Delta v_{s,d}^c - v_{c,q0}\Delta\theta,$$

$$\Delta v_{s,q}^s = \Delta v_{s,q}^c + v_{s,d0}\Delta\theta. \tag{6.30}$$

With mapping references of modules to the overall inputs, the connection matrices of the system are

$$R_{41} = \begin{bmatrix} O_{2\times6},I & O_{2\times3},Jv_{s,dq0},O_{2\times1} \\ O_{4\times10} & [0,0,1,0]^T \\ O_{2\times9} & -Jv_{g,dq0},O_{2\times1} \\ O_{2\times4},I & O_{2\times3},-Jv_{c,dq0},O_{2\times1} \\ I,O_{2\times7} & -Ji_{g,dq0},O_{2\times1} \\ -I,I & O_{2\times5},-Ji_{c,dq0},O_{2\times1} \\ O_{1\times8} & 1,0,0 \\ \begin{matrix} O_{4\times2} \\ 1.5I \\ O \\ 1.5I \\ O \end{matrix} & O_{12\times9} \end{bmatrix}, R_{42} = \begin{bmatrix} O_{2\times7} \\ I,O_{2\times5} \\ O_{1\times7} \\ I,O_{2\times5} \\ O_{7\times5} \\ O_{2\times4}, \begin{bmatrix} 0,1,0 \\ 1,0,0 \end{bmatrix} \\ 1.5I,O_{2\times5} \\ O_{2\times7} \\ 1.5I,O_{2\times5} \\ O_{2\times7} \\ O_{1\times6},1 \\ O_{1\times2},1,O_{1\times4} \end{bmatrix} \tag{6.31}$$

$$R_{43} = \begin{bmatrix} I,O_{2\times9} \end{bmatrix}, R_{44} = O_{2\times6}. \tag{6.32}$$

The coefficient matrices then are

$$A_{VSM} = \begin{bmatrix}
A_{VSM}^{1,1}I+A_{VSM}^{1,2}J & 0 & -\omega_B/L_sI & 0 & 0 & A_{VSM}^{1,11}I & A_{VSM}^{1,13}I & A_{VSM}^{1:2,15:16} & 0 \\
0 & A_{VSM}^{3,3}I+A_{VSM}^{1,2}J & \omega_B/L_gI & 0 & 0 & 0 & 0 & 0 & 0 \\
A_{VSM}^{5,1}I & -A_{VSM}^{5,1}I & A_{VSM}^{5,5}I+A_{VSM}^{1,2}J & 0 & 0 & 0 & 0 & 0 & 0 \\
0 & 0 & -I & 0 & 0 & 0 & 0 & A_{VSM}^{7:8,15:16} & \begin{bmatrix}0 & A_{VSM}^{7,18}\\0 & 0\end{bmatrix} \\
0 & (\beta_v-1)I & -k_{vp}I+B_{C_f}J & k_{vi}I & 0 & 0 & 0 & \begin{bmatrix}B_{vi}^{5,11} & A_{VSM}^{9,16}\\B_{vi}^{6,11} & A_{VSM}^{10,16}\end{bmatrix} & \begin{bmatrix}0 & A_{VSM}^{9,18}\\0 & 0\end{bmatrix} \\
R_{12}^{5,9}I & A_{VSM}^{11,3}I+X_fJ & A_{VSM}^{11,5}I+B_{C_f}k_{ip}J & k_{ip}k_{vi}I & k_{ii}I & A_{VSM}^{11,11}I & A_{VSM}^{11,13}I & \begin{bmatrix}B_{vi}^{7,11} & A_{VSM}^{11,16}\\B_{vi}^{8,11} & A_{VSM}^{12,16}\end{bmatrix} & \begin{bmatrix}0 & A_{VSM}^{11,18}\\0 & 0\end{bmatrix} \\
0 & 0 & 0 & 0 & 0 & I & 0 & 0 & 0 \\
0 & 0 & 0 & 0 & 0 & 0 & 0 & \begin{bmatrix}-K_D/T_a & 0\\\omega_B & 0\end{bmatrix} & \begin{bmatrix}-1/T_a & 0\\0 & 0\end{bmatrix} \\
0 & \begin{bmatrix}A_{VSM}^{17,3} & A_{VSM}^{17,4}\\A_{VSM}^{18,3} & A_{VSM}^{18,4}\end{bmatrix} & 0 & 0 & 0 & 0 & 0 & 0 & -\begin{bmatrix}\omega_{Pc} & 0\\0 & \omega_{Qc}\end{bmatrix}
\end{bmatrix},$$

where,

$$A_{VSM}^{1,1} = -R_s\omega_B/L_s,$$
$$A_{VSM}^{1,2} = -\omega_B\omega_{pu},$$
$$A_{VSM}^{3,3} = -R_g\omega_B/L_g,$$
$$A_{VSM}^{11,3} = K_{ad}\beta_{ad}+k_{ip}(\beta_v-1),$$
$$A_{VSM}^{17,3} = 1.5v_{g,d0}\omega_{Pc},$$
$$A_{VSM}^{17,4} = 1.5v_{g,q0}\omega_{Pc},$$
$$A_{VSM}^{18,3} = 1.5v_{g,q0}\omega_{Qc},$$
$$A_{VSM}^{18,4} = -1.5v_{g,d0}\omega_{Qc},$$
$$A_{VSM}^{5,5} = -G_{Cf}\omega_B/C_f,$$
$$A_{VSM}^{11,5} = -k_{ip}k_{vp},$$
$$A_{VSM}^{1,11} = -\frac{4\omega_B}{3L_sT_{sw}},$$
$$A_{VSM}^{11,11} = -\frac{8}{3T_{sw}},$$
$$A_{VSM}^{1,13} = -A_{VSM}^{1,11}/T_{sw},$$
$$A_{VSM}^{11,13} = -A_{VSM}^{11,11}/T_{sw},$$
$$A_{VSM}^{1:2,16} = \begin{bmatrix}\mathbf{0}_{2\times1} & (\omega_B/L_s)\mathbf{J}v_{s,dq0}\end{bmatrix},$$
$$A_{VSM}^{7:8,15:16} = \begin{bmatrix}\mathbf{0}_{2\times1} & \mathbf{J}v_{c,dq0}\end{bmatrix},$$
$$A_{VSM}^{9,16} = B_{C_f}v_{c,d0}+i_{g,q0}(\beta_v-1)-k_{vp}v_{c,q0},$$
$$A_{VSM}^{10,16} = B_{C_f}v_{c,q0}-i_{g,d0}(\beta_v-1)+k_{vp}v_{c,d0},$$
$$A_{VSM}^{11,16} = v_{v,q0}\beta_i+R_{12}^{5,9}i_{c,q0}+i_{g,q0}k_{ip}(\beta_v-1)+\omega_{pu}B_{vi}^{8,11}-k_{ip}k_{vp}v_{c,q0},$$
$$A_{VSM}^{12,16} = -v_{g,d,0}\beta_i-R_{12}^{5,9}i_{c,d0}-i_{g,d0}k_{ip}(\beta_v-1)-\omega_{pu}B_{vi}^{7,11}+k_{ip}k_{vp}v_{c,d0},$$
$$A_{VSM}^{7,18} = -K_q,$$
$$A_{VSM}^{9,18} = -K_qk_{vp},$$
$$A_{VSM}^{11,18} = A_{VSM}^{9,18}k_{ip}.$$

$$
B_{VSM} = \begin{bmatrix}
O_{2\times7} \\
-\omega/L_g I, O_{2\times5} \\
O_{2\times7} \\
O, I, O, \left[K_q, 0 \right]^T \\
O, k_{vp} I, O, \left[K_q k_{vp}, 0 \right]^T \\
\beta_i I, k_{ip} k_{vp} I, O, \left[K_q k_{ip} k_{vp}, 0 \right]^T \\
O_{2\times7} \\
O_{2\times4}, 1/T_a \begin{bmatrix} K_D & 1 \\ 0 & 0 \end{bmatrix}, O \\
\dfrac{3}{2} \begin{bmatrix} i_{g,d0}\omega_{Pc} & i_{g,q0}\omega_{Pc} \\ -i_{g,q0}\omega_{Qc} & i_{g,d0}\omega_{Qc} \end{bmatrix}, O_{2\times5}
\end{bmatrix}
\tag{6.34}
$$

$$
C_{VSM} = [O, I, O_{2\times14}], D_{VSM} = O_{2\times7}.
$$

6.7.1.1   Time Domain Validation

The circuit and controller parameters of the studied VSM are listed in Table 6.1. Based on the provided parameters, the system is given a disturbance of 0.02 pu on the power reference magnitude from 0.70 to 0.72 pu. Figure 6.13 presents the responses of controlled grid currents $\mathbf{i}_{dq}$ in both simulation and established state-space model through equations (6.33) and (6.34). It can be seen that both models present approximate transient responses. The correctness of the derived state-space model is thus verified.

6.8   DERIVATION AND VALIDATION OF THE ADMITTANCE MODEL

The admittance model of the derived SSM is determined by the multiple-input-multiple-output matrices in $dq-$ frame as

$$
\begin{bmatrix} Y_{dd} & Y_{dq} \\ Y_{qd} & Y_{qq} \end{bmatrix} \begin{bmatrix} v_d \\ v_q \end{bmatrix} = \begin{bmatrix} i_d \\ i_q \end{bmatrix},
\tag{6.35}
$$

or $\mathbf{Y}\mathbf{v}_{dq} = \mathbf{i}_{dq}$ in compact form.

TABLE 6.1
Circuit and Controller Parameters of the Studied virtual synchronous generator (VSG)

Parameter	Value
DC-link voltage V_{dc}	2,000 V
Grid fundamental frequency f_1	50 Hz
Electrical angular base frequency ω_B	100π rad/s
Three-phase voltage base V_{Base}	690 V
Power base S_{Base}	10 kVA
Filter inductance L_s	0.3 pu
Filter resistance R_s	0.06 pu
Grid inductance L_g	0.7 pu
Grid resistance R_g	0.14 pu
Filtering capacitor C_f	0.005 pu
Paralleled filter conductance G_{Cf}	0.01 pu
Switching period T_{sw}	0.1 ms
Proportional gain of current controller K_{ip}	0.0032 pu
Integral gain of current controller K_{ii}	0.20 pu/s
Closed-loop bandwidth of current controller ω_{ic}	1250 Ω
Proportional gain of voltage controller K_{vp}	1 pu
Integral gain of voltage controller K_{vi}	416.67 pu/s
Closed-loop bandwidth of inner voltage controller ω_{vc}	0.0066 Ω
Current feedforward coefficient β_i	0.5
Voltage feedforward coefficient β_v	0.5
Active damping coefficient β_{ad}	0
Power swinging time constant T_a	2 s
Angular speed deviation damping coefficient K_D	1,000 pu
Reactive power droop coefficient K_q	0
Active power LPF bandwidth ω_{Pc}	10π rad/s
Reactive power LPF bandwidth ω_{Qc}	10π rad/s

From the derived SSM of VSM, the response of PCC currents with respect to voltages is determined by the control theory that

$$\Delta H(s) = \frac{\Delta y_{VSM}(s)}{\Delta u_{VSM}(s)} = C_{VSM}\left(sI_{18\times18} - A_{VSM}\right)^{-1} B_{VSM} + D_{VSM}. \qquad (6.36)$$

The admittance matrix is then formed as

$$Y(s) = -\begin{bmatrix} H_{1,1}(s) & H_{1,2}(s) \\ H_{2,1}(s) & H_{2,2}(s) \end{bmatrix} = -\begin{bmatrix} i_d/v_d & i_d/v_q \\ i_q/v_d & i_q/v_q \end{bmatrix}, \qquad (6.37)$$

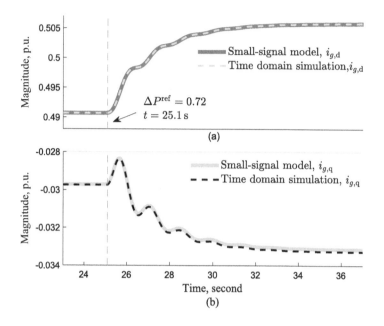

FIGURE 6.13 Time-domain validation of established VSM model by observing $\mathbf{i}_{g,dq}$ under converter's synchronous frame after the power reference step change of 0.02 pu.

by inserting a series of frequencies where $s = j\omega$. Negative values result from the defined current direction by determining admittance toward VSM. Figure 6.14 compares the derived analytical admittance model of VSM with the simulation at PCC. The two models approximately match each other. Varying power reference set-points from 0.1 to 0.7 pu does not vary the admittance shape of VSM significantly. The figure also reveals the instability zone at 1,573 Hz where the angle of admittance jumps across 180° on $dd-$ and $qq-$ axes. However, the VSM shows passivity in a wide range from 1 Hz till 1,573 Hz under this parameter sets.

Figure 6.15 varies the closed-loop bandwidth of inner current controller from 0.05 times of the original setting to 1.5 times. The very low bandwidth shapes the low-frequency characteristics of VSM showing capacitance feature while gradually moving to inductance feature. The admittance magnitude increases drastically in the low-frequency region.

Figure 6.16 presents impacts on voltage closed-loop bandwidth from 0.4 time of the previous setting to 1.78 times, while the converter works at power-voltage mode sending 0.7 pu power. At 1 Hz, the extremely low bandwidth forces VSM to show resistivity. Admittance magnitude decreases with the increased bandwidth.

Figure 6.17 further investigates poles of the system presented in Figure 6.16. The voltage closed-loop bandwidth keeps system stable when it is between 0.2 and 1.9 times of initial setting. The low-frequency poles gradually disappear when increasing the bandwidth. However, with increased bandwidth, it further brings the system back to instability. This makes sense since the voltage control bandwidth is larger than the outer power loop bandwidth and smaller than the inner current loop bandwidth.

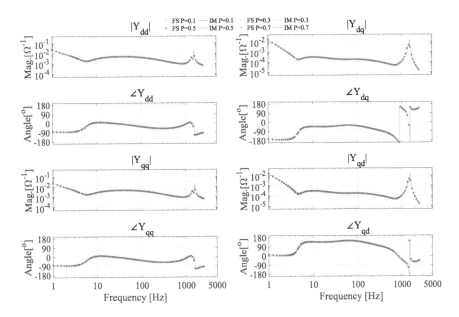

FIGURE 6.14 Admittance validation of established VSM model between analytical model and simulation, varying active power reference from 0.1 to 0.7 pu with 0.2 pu step. The VSM works in power-voltage (PV) mode.

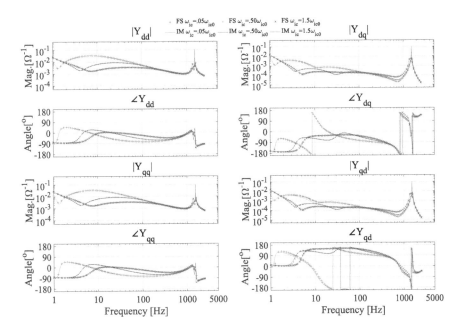

FIGURE 6.15 Admittance validation of established VSM model between analytical model and simulation, varying current control bandwidth at the ratios of 0.01, 0.5, and 1.5 times. The VSM works in PV mode.

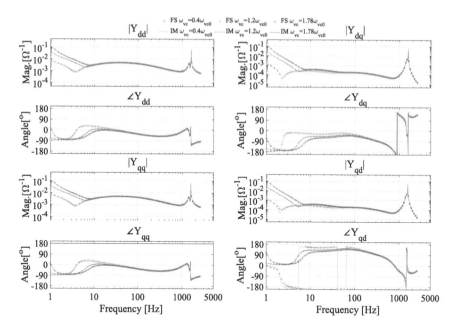

FIGURE 6.16 Admittance validation of established VSM model between analytical model and simulation, varying voltage control bandwidth at the ratios of 0.4, 1.2, and 1.78 times. The VSM works in PV mode.

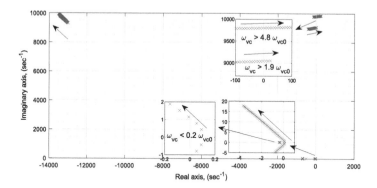

FIGURE 6.17 The poles of the dd- and qq-axes elements of VSM by varying the close-loop voltage control bandwidth from 0.10 to 5.0 time of the original bandwidth, while keeping other parameters constant.

Figure 6.18 presents the admittance behavior of VSM under the same condition as that in Figure 6.16 but working at active power-reactive power mode, where $K_q = 0.2$. The voltage closed-loop bandwidth varies from 0.5 to 5 times of original setting. At a very large bandwidth, VSM shows significant capacitance feature below 5 Hz. Lower bandwidth setup shows opposite magnitude feature compared to higher settings.

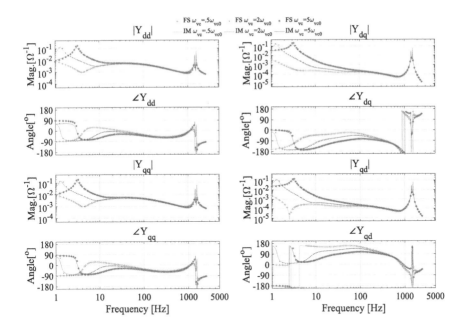

FIGURE 6.18 Admittance validation of established VSM model between analytical model and simulation, varying voltage control bandwidth at the ratios of 0.5, 2.0, and 5.0 times. The VSM works in real-reactive power (PQ) mode.

6.9 CONCLUSION

In this chapter, the state-space model of the VSM has been built, and its admittance model has consequently been derived. The correctness of the derived models has been verified in both time domain and frequency domain. In the future, the derived state-space and admittance models can be used for various applications, e.g., controller design, converter-grid interaction, eigenvalue analysis, and impedance reshaping. In the future study, the small-signal models of other grid-forming control strategies can be derived in the similar way.

REFERENCES

[1] Benjamin Kroposki, Brian Johnson, Yingchen Zhang, Vahan Gevorgian, Paul Denholm, Bri-Mathias Hodge, and Bryan Hannegan. Achieving a 100% renewable grid: Operating electric power systems with extremely high levels of variable renewable energy. *IEEE Power and Energy Magazine*, 15(2): 61–73, 2017.
[2] Federico Milano, Florian Dörfler, Gabriela Hug, David J Hill, and Gregor Verbič. Foundations and challenges of low-inertia systems. In *2018 Power Systems Computation Conference (PSCC)*, 1–25, IEEE, Dublin, 2018.
[3] Mario Paolone, Trevor Gaunt, Xavier Guillaud, Marco Liserre, Sakis Meliopoulos, Antonello Monti, Thierry Van Cutsem, Vijay Vittal, and Costas Vournas. Fundamentals of power systems modelling in the presence of converter-interfaced generation. *Electric Power Systems Research*, 189: 106811, 2020.

[4] Robert H Lasseter, Zhe Chen, and Dinesh Pattabiraman. Grid-forming inverters: A critical asset for the power grid. *IEEE Journal of Emerging and Selected Topics in Power Electronics*, 8(2): 925–935, 2019.

[5] Dinesh Pattabiraman, RH Lasseter, and TM Jahns. Comparison of grid following and grid forming control for a high inverter penetration power system. In *2018 IEEE Power & Energy Society General Meeting (PESGM)*, 1–5. IEEE, Portland, OR, 2018.

[6] Lizhi Ding, Yuxi Men, Yuhua Du, Xiaonan Lu, Bo Chen, Jin Tan, and Yuzhang Lin. Region-based stability analysis of resilient distribution systems with hybrid grid-forming and grid-following inverters. In *2020 IEEE Energy Conversion Congress and Exposition (ECCE)*, 3733–3740. IEEE, Detroit, MI, 2020.

[7] Brian J Pierre, Hugo N Villegas Pico, Ryan T Elliott, Jack Flicker, Yashen Lin, Brian B Johnson, Joseph H Eto, Robert H Lasseter, and Abraham Ellis. Bulk power system dynamics with varying levels of synchronous generators and grid-forming power inverters. In *2019 IEEE 46th Photovoltaic Specialists Conference (PVSC)*, 0880–0886. IEEE, Chicago, IL, 2019.

[8] Mohammed Masum Siraj Khan, Yashen Lin, Brian Johnson, Mohit Sinha, and Sairaj Dhople. Stability assessment of a system comprising a single machine and a virtual oscillator controlled inverter with scalable ratings. In *IECON 2018–44th Annual Conference of the IEEE Industrial Electronics Society*, 4057–4062. IEEE, Washington, DC, 2018.

[9] Yashen Lin, Brian Johnson, Vahan Gevorgian, Victor Purba, and Sairaj Dhople. Stability assessment of a system comprising a single machine and inverter with scalable ratings. In *2017 North American Power Symposium (NAPS)*, 1–6. IEEE, Morgantown, WV, 2017.

[10] Wenhua Wu, Yandong Chen, Leming Zhou, An Luo, Xiaoping Zhou, Zhixing He, Ling Yang, Zhiwei Xie, Jinming Liu, and Mingmin Zhang. Sequence impedance modeling and stability comparative analysis of voltage controlled VSGS and current-controlled VSGS. *IEEE Transactions on Industrial Electronics*, 66(8): 6460–6472, 2018.

[11] Yunyang Xu, Heng Nian, Yangming Wang, and Dan Sun. Impedance modelling and stability analysis of VSG controlled grid-connected converters with cascaded inner control loop. *Energies*, 13(19): 5114, 2020.

[12] Yunyang Xu, Heng Nian, Bin Hu, and Dan Sun. Impedance modelling and stability analysis of VSG controlled type-iv wind turbine system. *IEEE Transactions on Energy Conversion*, 36(4): 3438–3448, 2021.

[13] G. Gaba, S. Lefebvre, and D. Mukhedkar. Comparative analysis and study of the dynamic stability of AC/DC systems. *IEEE Transactions Power Systems*, 3(3):978–985, 1988.

[14] Prabha S Kundur, Neal J Balu, and Mark G Lauby. Power system dynamics and stability. *Power System Stability and Control*, 3, 2017.

[15] Amirnaser Yazdani and Reza Iravani. *Voltage-Sourced Converters in Power Systems: Modeling, Control, and Applications*. John Wiley & Sons, Hoboken, NJ, 2010.

[16] Taoufik Qoria, François Gruson, Frédéric Colas, Xavier Guillaud, Marie-Sophie Debry, and Thibault Prevost. Tuning of cascaded controllers for robust grid-forming voltage source converter. In *2018 Power Systems Computation Conference (PSCC)*, 1–7. IEEE, Dublin, 2018.

7 Grid-Forming Control of Doubly Fed Induction Generators

Santiago Arnaltes, José Luis Rodríguez Amenedo, and Jesús Castro
University Carlos III of Madrid

CONTENTS

7.1   INTRODUCTION

Over the last decades, wind power generation has been the source of the renewable energy that has experienced the largest growth around the world. By the end of the last century, most of wind turbines used fixed-speed induction generators. But, the development of more demanding grid code requirements gave a significant advantage to variable-speed wind turbines. Nowadays, two variable-speed wind turbine generator technologies dominate the market: full-converter generators and doubly fed induction generators (DFIGs). Full-converter generators deliver all the power generated by the wind turbine to the grid through a power electronic converter. This configuration decouples the generator voltage and frequency from the grid, allowing variable-speed operation and preventing grid disturbances, such as voltage dips,

DOI: 10.1201/9781003302520-7

affecting the generator. In DFIGs, the stator is directly connected to the grid, while the rotor windings are fed at variable frequency through a back-to-back converter. This allows variable-speed operation of the wind turbine, while requiring a smaller converter compared to full-converter wind turbines. The converter exchanges only the generator slip power, which usually does not exceed 30% of the nominal power of the wind turbine, being this the main advantage of this system. Traditionally, stator-flux-oriented vector control [1], direct torque control [2], or model predictive control [3] have been employed to regulate the generator active and reactive power. Since these control techniques maintain constant power, following commanded set-points, regardless of the grid disturbances, they are referred as grid following.

Nevertheless, electrical grids are progressively being decarbonized through the increasing penetration of renewable energies, most of which use power electronic converter interfaces. These technologies are displacing conventional synchronous generators (SGs) that are key to ensure the grid stability. Therefore, new requirements are established for power electronic-based generators to provide grid-supporting services similar to that of SG, so the stability and robustness of the grid is not compromised. It has been demonstrated that system stability and robust operation can be guaranteed in the absence of SG, provided that enough power electronic-based generation exhibits grid-forming (GFM) capabilities [4,5], because of their inherent inertia provision. Although some studies have proposed the provision of inertia through the tuning of the phase-locked loop dynamics [6,7], DFIGs can also be controlled as virtual synchronous machines (VSMs) to support the grid voltage and frequency stability. In [8], a VSM is implemented in a DFIG, using an active power synchronization loop to obtaining directly the rotor voltage angle. However, in a DFIG, controlling directly the rotor voltage does not lead to the control of a voltage source behind an impedance, which is the principle of VSM. On the other hand, due to the fact that GFM technologies behave as voltage sources, their low voltage ride-through capability becomes an issue, because they naturally respond to faults with high short-circuit currents, that cannot be withstand by power electronics converters. In [9], an improved VSM control scheme for a DFIG is proposed for fault ride-through.

In this chapter, a GFM control scheme for DFIGs is proposed, so they behave as real voltage sources. The proposed GFM control is based on the rotor flux orientation to a reference axis obtained from the emulation of the SG swing equation. The rotor flux is then oriented to the reference axis by means of a flux controller that also controls the flux magnitude. The flux orientation in turn allows to control the DFIG torque, while the flux magnitude control allows to regulate the generator reactive power or terminal voltage. The proposed control system has been validated through a comprehensive simulation, assessing its GFM capability. Moreover, the small signal analysis has also been performed to assess the system stability.

7.2 SYSTEM DESCRIPTION

Figure 7.1 shows the overall scheme of the DFIG GFM control. The DFIG consists of an induction generator with the stator directly connected to the grid and the rotor fed at variable frequency through a back-to-back converter. The back-to-back converter

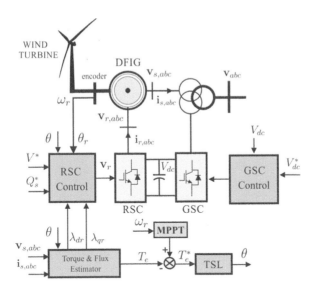

FIGURE 7.1   DFIG grid-forming control scheme.

is made of two converters with a common DC link. One converter connected to the rotor (RSC) and the other connected to the grid (GSC).

The DFIG control is conducted as follows. The torque synchronization loop (TSL) block determines the control angle θ, which determines the position of the d-axis of the rotary reference system. The input of this block is the error between the reference torque, T_e^*, and the estimated electromagnetic torque T_e. The reference torque is calculated by the wind turbine controller based on the turbine rotational speed, following a maximum power point tracking (MPPT) strategy at partial load and keeping it constant at full load.

The stator voltage, $v_{s,abc}$, and the stator and rotor currents, $i_{s,abc}$ and $i_{r,abc}$, are measured and used for the estimation of the DFIG electromagnetic torque and rotor flux. Then, using the control angle θ, the dq components of the rotor flux vector, λ_{dr} and λ_{qr}, are calculated. The RSC control is based on the regulation of the dq components of the rotor flux. The component λ_{qr}^* is regulated to 0 so that the rotor flux vector is aligned with the d-axis of the rotary reference frame, and the component λ_{dr}^* is calculated from a voltage regulator (or reactive power regulator). The error between these reference values and the estimated ones, λ_{dr} and λ_{qr}, allows the flux regulators to calculate the rotor voltage components, v_{dr} and v_{qr}. Then, the instantaneous rotor voltages $v_{r,abc}$ are calculated by applying the coordinate transformation using the control angle θ and the rotor position θ_r measured by the encoder coupled to the DFIG shaft.

The GSC can be controlled in a conventional way, by maintaining constant DC bus voltage, V_{dc}, using voltage-oriented control [10], so the rotor power is instantaneously drawn or injected to the grid, depending on the DFIG slip sign.

7.3   DFIG DYNAMIC MODEL

The electrical equations of an induction machine in a reference frame rotating at a speed of ω can be expressed as

$$\vec{v}_s = -R_s \vec{i}_s + \frac{d\vec{\lambda}_s}{dt} + j\omega\vec{\lambda}_s \tag{7.1}$$

$$\vec{v}_r = R_r \vec{i}_r + \frac{d\vec{\lambda}_r}{dt} + j\left(\lambda - \omega_r\right)\vec{\lambda}_r$$

where $\vec{v}_s$, $\vec{v}_r$, $\vec{i}_s$, $\vec{i}_r$, $\vec{\lambda}_s$, $\vec{\lambda}_r$ are the stator and rotor voltage, current, and flux vectors, respectively, R_s, R_r, the stator and rotor resistances, and ω_r, the rotor electrical rotational speed.

The relationship between fluxes and currents is the following:

$$\vec{\lambda}_s = -L_s \vec{i}_s + L_m \vec{i}_r \tag{7.2}$$

$$\vec{\lambda}_r = -L_m \vec{i}_s + L_r \vec{i}_r$$

where L_s and L_r are the stator and rotor inductances and L_m the magnetizing inductance.

By solving equation (7.2) for the current vectors as a function of the flux vectors, the following equations are obtained:

$$\vec{i}_s = \frac{-1}{\sigma L_s}\left(\vec{\lambda}_s - \frac{L_m}{L_r}\vec{\lambda}_r\right) \tag{7.3}$$

$$\vec{i}_r = \frac{1}{\sigma L_r}\left(\vec{\lambda}_r - \frac{L_m}{L_s}\vec{\lambda}_s\right)$$

where $\sigma = 1-\left(L_m^2 / (L_s L_r)\right)$ is the leakage coefficient.

By substituting equation (7.3) in equation (7.1), the dynamic DFIG equations are obtained, being the stator and rotor flux vectors, $\vec{\lambda}_s$, $\vec{\lambda}_r$, the state variables, and the stator and rotor voltage vectors, $\vec{v}_s$, $\vec{v}_r$, the inputs. These equations are as follows:

$$\vec{v}_s = \frac{1}{\sigma T_s}\left(\vec{\lambda}_s - \frac{L_m}{L_r}\vec{\lambda}_r\right) + \frac{d\vec{\lambda}_s}{dt} + j\omega\vec{\lambda}_s \tag{7.4}$$

$$\vec{v}_r = \frac{1}{\sigma T_r}\left(\vec{\lambda}_r - \frac{L_m}{L_s}\vec{\lambda}_s\right) + \frac{d\vec{\lambda}_r}{dt} + j(\omega - \omega_r)\vec{\lambda}_r$$

where $T_s = L_s/R_s$ and $T_r = L_r/R_r$ are the stator and rotor time constants.

Finally, regrouping terms in equation (7.4), the following equations, representing the DFIG dynamic model, are obtained:

$$\frac{d\vec{\lambda}_s}{dt} = -\left(\frac{1}{\sigma T_s} + j\omega\right)\vec{\lambda}_s + \frac{1}{\sigma T_s}\frac{L_m}{L_r}\vec{\lambda}_r + \vec{v}_s \tag{7.5}$$

$$\frac{d\vec{\lambda}_r}{dt} = -\left(\frac{1}{\sigma T_r} + js\omega\right)\vec{\lambda}_r + \frac{1}{\sigma T_r}\frac{L_m}{L_s}\vec{\lambda}_s + \vec{v}_r$$

where $s = (\omega - \omega_r)/\omega$ is the slip.

7.4   DFIG EQUIVALENT CIRCUIT

The electrical equation of the stator (equation 7.1) in steady state, $d\vec{\lambda}_s/dt = 0$, is as follows:

$$\vec{v}_s = -R_s\vec{i}_s + j\omega\vec{\lambda}_s \tag{7.6}$$

By substituting in this equation, the stator flux vector, $\vec{\lambda}_s$, as a function of $\vec{\lambda}_r$ and $\vec{i}_s$, according to equation (7.3),

$$\vec{\lambda}_s = -\sigma L_s\vec{i}_s + \frac{L_m}{L_r}\vec{\lambda}_r \tag{7.7}$$

the following equation is obtained:

$$\vec{v}_s = -\left(R_s + j\sigma X_s\right)\vec{i}_s + j\omega\frac{L_m}{L_r}\vec{\lambda}_r \tag{7.8}$$

where $X_s = \omega L_s$ is the stator reactance.

By defining the DFIG internal electromotive force (e.m.f.) as

$$\vec{e}_s = j\omega\frac{L_m}{L_r}\vec{\lambda}_r \tag{7.9}$$

the electrical equation of the stator can be expressed as a voltage source behind an impedance

$$\vec{e}_s = \left(R_s + j\sigma X_s\right)\vec{i}_s + \vec{v}_s \tag{7.10}$$

Figure 7.2 shows the DFIG equivalent circuit according to this equation. This circuit is analog to the equivalent circuit of an SG, being the reactance σX_s analog to the synchronous reactance of the SG. However, based on the electrical parameters of Appendix A, $\sigma X_s = 0.23$ pu, which is lower than the SG synchronous reactance whose value is typically around 1 pu. Moreover, in the SG, the internal e.m.f., is a function of the rotor flux, generated by the excitation current. In the DFIG, as shown

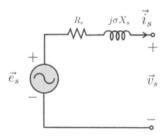

FIGURE 7.2 DFIG stator equivalent circuit.

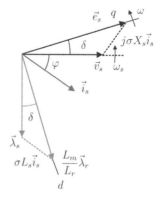

FIGURE 7.3 DFIG vector diagram.

in equation (7.10), $\vec{e}_s$ also depends on the rotor flux, which can be controlled directly by the voltage applied to the rotor $\vec{v}_r$, as given in the second equation of equation (7.5).

In Figure 7.3, the vector diagram of a DFIG is presented. The stator voltage vector $\vec{v}_s$ is taken as phase reference and the current $\vec{i}_s$ is lagging by an angle φ, which means that the DFIG is delivering active and reactive power to the grid. The stator flux vector $\vec{\lambda}_s$ is lagging an angle of 90°, $\vec{v}_s \sim j\omega\vec{\lambda}_s$, and the rotor flux vector $\vec{\lambda}_r$ can be calculated from equation (7.3) as

$$\vec{\lambda}_r = \frac{L_r}{L_m}(\vec{\lambda}_s + \sigma L_s \vec{i}_s) \tag{7.11}$$

which, as shown in the vector diagram, leads the stator flux vector by an angle of δ. According to equation (7.10), the stator e.m.f. $\vec{e}_s$ leads 90° over $\vec{\lambda}_r$ and leads the same angle of δ over the stator voltage vector. As shown in the next section, δ is the so-called load or torque angle.

By defining a reference frame with a d-axis aligned with $\vec{\lambda}_r$ and a q-axis aligned with $\vec{e}_s$, the analogy of the DFIG with a SG is complete. The excitation current generates the rotor flux in the direct axis, inducing an e.m.f. in the quadrature axis.

7.5 DFIG ELECTROMAGNETIC TORQUE AND REACTIVE POWER

The DFIG stator complex power as a function of vectors $\vec{v}_s$ and $\vec{i}_s$ is equal to

$$\vec{S}_s = P_s + jQ_s = \frac{3}{2}\vec{v}_s\left(\vec{i}_s\right)^{\dagger} \tag{7.12}$$

where the operator (†) indicates conjugated complex. In steady state, neglecting the stator resistance, $\vec{v}_s \sim j\omega\vec{\lambda}_s$. Then, the expressions of the active and reactive power generated at the DFIG stator are as follows:

$$P_s = R_e\left\{\frac{3}{2}\left(j\omega\vec{\lambda}_s\right)\left(\vec{i}_s\right)^{\dagger}\right\} = \frac{3}{2}\omega I_m\left\{\left(\vec{\lambda}_s\right)^{\dagger}\vec{i}_s\right\} \tag{7.13}$$

$$Q_s = I_m\left\{\frac{3}{2}\left(j\omega\vec{\lambda}_s\right)\left(\vec{i}_s\right)^{\dagger}\right\} = \frac{3}{2}\omega R_e\left\{\vec{\lambda}_s\left(\vec{i}_s\right)^{\dagger}\right\} \tag{7.13}$$

Having neglected R_s, the stator power P_s can be regarded as the air gap power. Then,

$$T_e = \frac{P_s}{\frac{\omega}{p}} = \frac{3}{2}pI_m\left\{\left(\vec{\lambda}_s\right)^{\dagger}\vec{i}_s\right\} \tag{7.14}$$

where p is the number of pole pairs.

From equation (7.4), $\left(\vec{\lambda}_s\right)^{\dagger}\vec{i}_s$ can be written as

$$\left(\vec{\lambda}_s\right)^{\dagger}\vec{i}_s = \frac{-1}{\sigma L_s}\left(\lambda_s^2 - \frac{L_m}{L_r}\left(\vec{\lambda}_s\right)^{\dagger}\vec{\lambda}_r\right) \tag{7.15}$$

And by substituting in equations (7.14) and the second equation of (7.13), yields to

$$T_e = \frac{3}{2}p\left(\frac{L_m}{\sigma L_s L_r}\right)\lambda_s\lambda_r\sin\delta \tag{7.16}$$

$$Q_s = \frac{3}{2}\left(\frac{\omega}{\sigma L_s}\right)\lambda_s\left(\frac{L_m}{L_r}\lambda_r\cos\delta - \lambda_s\right)$$

where δ is the angular difference between the stator and rotor flux vectors, the so-called torque angle. These expressions show that the DFIG torque can be controlled by controlling the torque angle, while the DFIG reactive power can be controlled by controlling the rotor flux magnitude, in a similar way as an SG.

Considering the DFIG internal e.m.f. as defined in equation (7.9), the former equations can also be expressed as

$$T_e = \frac{3}{2} \frac{p}{\omega} \left(\frac{1}{\sigma X_s} \right) v_s e_s \sin\delta \tag{7.17}$$

$$Q_s = \frac{3}{2} \left(\frac{1}{\sigma X_s} \right) v_s \left(e_s \cos\delta - v_s \right)$$

These expressions are analog to those of the classical formulation for the active and reactive power supplied by an SG connected to an infinite power bus, and they could have been obtained easily from the equivalent circuit of Figure 7.2, as well.

7.6 DFIG TORQUE SYNCHRONIZATION LOOP

As illustrated in Figure 7.1, the TSL block sets the control angle θ that determines the position of the d-axis, relative to a stationary reference system, based on the difference between the reference torque T_e^* and the actual torque T_e. The rotational speed of the d-axis is ω (as shown in Figure 7.3), and it is calculated as follows

$$\omega = \frac{1}{J} \int \left(T_e^* - T_e - D(\omega - \omega_0) \right) dt \tag{7.18}$$

where ω_0 is the reference angular frequency, J is the emulated inertia constant, in seconds, and D is the damping constant, being its inverse $R = 1/D$ the droop constant, which is the ratio between the normalized frequency variation, $\Delta f / f_n$, and the normalized power deviation, $\Delta P / P_n$. The control angle θ is then obtained by the integration of ω, as shown in Figure 7.4. As it will be shown in the next paragraph, the rotor flux will be oriented to this angular position, while θ is continuedly adjusted by the TSL so that the angular difference with the stator flux is the required torque angle.

The block diagram presented in Figure 7.4 denotes the TSL implementation, but it does not explain the synchronizing principles. For such purposes, the system is

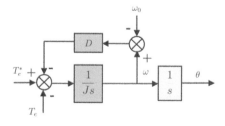

FIGURE 7.4   Torque synchronization loop.

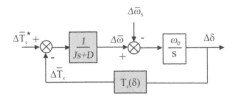

FIGURE 7.5   Linearized TSL block diagram.

linearized around an equilibrium point. The angular increment $\Delta\delta$ is calculated based on the following equation:

$$\frac{1}{\omega_0}\frac{d\Delta\delta}{dt} = \left(\Delta\bar{\omega} - \Delta\bar{\omega}_s\right) \tag{7.19}$$

where $\Delta\bar{\omega}$ and $\Delta\bar{\omega}_s$ are the angular frequency increments of the dq-axes system and the stator flux in pu.

As shown in Figure 7.5, the torque increment $\Delta\bar{T}_e$ depends on $\Delta\delta$ as the function $T_e(\delta)$. The first equation of (7.16) denotes a non-linear relationship between torque T_e and angle δ. Considering increments in this equation, the following linear relationship between torque and angle is obtained

$$\Delta\bar{T}_e = K_s\Delta\delta \tag{7.20}$$

where K_s is the synchronizing constant, calculated by applying the partial derivative of $\bar{T}_e$ with respect to δ in equation (7.17) at the equilibrium point "0".

$$K_s = \left(\frac{\partial\bar{T}_e}{\partial\delta}\right)_0 = \left(\frac{L_m}{\sigma L_s L_r}\right)\lambda_{s0}\lambda_{r0}\cos\delta_0 \tag{7.21}$$

The $(3/2)p$ constant does not appear in this equation, since torque $\bar{T}_e$ is expressed in pu.

By observing the block diagram of Figure 7.5 and considering the synchronizing constant K_s, the transfer function between $\Delta\delta$ and the reference torque $\Delta\bar{T}_e^*$ is

$$\frac{\Delta\delta}{\Delta\bar{T}_e^*} = \frac{\omega_0}{Js^2 + Ds + K_s\omega_0} \tag{7.22}$$

which is analog to the SG oscillation equation. This equation can be written as

$$\frac{J}{\omega_0}s^2\left(\Delta\delta\right) + \frac{D}{\omega_0}s\left(\Delta\delta\right) + K_s\left(\Delta\delta\right) = \Delta\bar{T}_e^* \tag{7.23}$$

By comparing the terms of equation (7.23) with the normalized expression of a second-order function $s^2 + 2\xi\omega_n s + \omega_n^2$, it is obtained that

$$\omega_n = \sqrt{\frac{\omega_o K_s}{J}} \tag{7.24}$$

$$\xi = \frac{D}{2\sqrt{J\omega_o K_s}}$$

where ω_n is the undamped natural frequency, and ξ is the damping coefficient.

In Figure 7.5, the DFIG torque depends on the reference torque, T_e^*, as well as on the stator frequency, ω_s. The election of the parameters D and J is not arbitrary. The damping constant D indicates the torque variation that the DFIG has to produce to maintain the system synchronized when the frequency changes. For a typical value of $R = 0.05$ pu., the damping constant is $D = 20$ pu, which denotes that if there is a frequency variation of 0.05 pu (2.5 Hz in a 50 Hz grid), the DFIG would have to increase its torque by 1 pu. On the other hand, if in equation (7.23) $J = 0$, the response of the DFIG torque to the inputs, reference torque or frequency, corresponds to a first-order function with a negative pole located on the real axis

$$\sigma = -\frac{K_s \omega_0}{D} \tag{7.25}$$

The smaller is D, the further the pole will be from the origin. Therefore, the system response will be faster and also more sensitive to frequency changes. Since $K_s\omega_0$ is a constant, the coefficient D is the first choice when determining the dynamics of the TSL. If the inertia constant J is gradually increased from 0, the loci of the system poles are presented in Figure 7.6. Initially, the system is overdamped ($\xi > 1$), and the poles are real negative until they reach the same value at -2σ, where $\xi = 1$. From this point, when J increases, the poles become conjugated complex, following a circumference path, as shown in Figure 7.6. It can be observed that this circumference has a radius of σ and the center is at $(-\sigma, 0)$. When the inertia constant increases, holding the value of D constant, the system damping decreases.

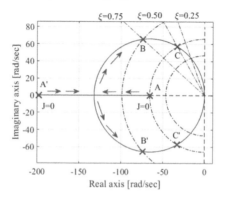

FIGURE 7.6   Pole loci under variation of J for $D = 20$ pu.

In Figure 7.7, the torque ΔT_e response is presented for a step of the reference input $\Delta \overline{T}_e^*$ from 0 to 1 pu at $t = 0.1$ s, for various values of J being $D = 20$ pu. The torque response for $J = 0$ is that of a first-order system with its pole located at A. The other cases analyzed are those with poles at BB' ($J=0.13$ s and $\xi = 0.75$) and CC' ($J=0.30$ s and $\xi = 0.50$). The unit step response shows the same conclusions in the time domain. When the inertia constant is increased, the response becomes slower and less damped.

In Figure 7.8, the response of the DFIG torque is presented for a frequency step from 0 to −0.05 pu, at $t=0.1$ s. The response reaches 1 pu, since the frequency step is equal to 1/D in magnitude in this case. However, the conclusions with respect to the dynamic response are similar to the previous case, being the overshoot higher for increasing values of the inertia constant.

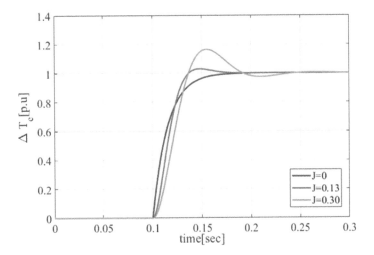

FIGURE 7.7   Torque response to a unit step reference.

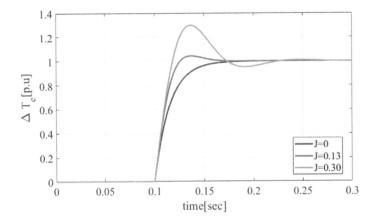

FIGURE 7.8   Torque response to a stator frequency step of −0.05 pu.

7.7   DFIG ROTOR FLUX CONTROL

The RSC control aims to regulate the electromagnetic torque and reactive power at the stator output. The reference electromagnetic torque T_e^* is calculated based on the wind turbine MPPT strategy (see Figure 7.1). This reference is compared to the actual DFIG torque, and as specified in the previous section, the synchronization loop calculates the control angle θ.

Torque can be easily estimated from the direct measurement of the stator power based on equation (7.14), using the instantaneous stator voltage and current measurements, $v_{s,abc}$ and $i_{s,abc}$.

$$T_e = \frac{3}{2}\frac{p}{\omega}\left(u_{\alpha s}i_{\alpha s} + u_{\beta s}i_{\beta s}\right) \tag{7.26}$$

Rotor flux can also be easily calculated from the direct measurements of the stator and rotor currents, $i_{s,abc}$ and $i_{r,abc}$. Then, the rotor flux vector in a stationary reference frame is calculated as

$$\vec{\lambda}_r^{\alpha\beta} = \lambda_{\alpha r} + j\lambda_{\beta r} = -L_m\vec{i}_s + L_r\vec{i}_r e^{j\theta r} \tag{7.27}$$

where θ_r is the rotor angular position, as measured by the encoder attached to the rotor shaft.

The rotor flux vector $\lambda_{dr} + j\lambda_{qr}$ in the rotating reference system is obtained by multiplying the vector $\lambda_{\alpha r} + j\lambda_{\beta r}$ by $e^{-j\theta}$.

$$\lambda_{dr} = \lambda_{\alpha r}\cos\theta + \lambda_{\beta r}\sin\theta \tag{7.28}$$

$$\lambda_{qr} = -\lambda_{\alpha r}\sin\theta + \lambda_{\beta r}\cos\theta$$

After calculating the rotor flux components in dq-axes, the RSC control is applied to maintain the rotor flux vector oriented along the d-axis of the rotating reference frame. In Figure 7.9, the position of the stator and rotor flux vectors, the dq-axes rotating at a speed of ω, as well as the torque angle δ and the control angle θ are shown.

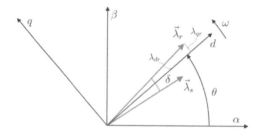

FIGURE 7.9   Stator and rotor flux vector diagram referred to dq-axes.

When the rotor flux vector, $\vec{\lambda}_r$, is aligned with the d-axis, the DFIG is operating in synchronism. To produce this alignment, it is necessary to control $\vec{\lambda}_r$ by means of the rotor voltage, $\vec{v}_r$. The relationship between both vectors, based on the second equation of (7.1) and disregarding the voltage drop at R_r, is as follows:

$$\vec{v}_r = \frac{d\vec{\lambda}_r}{dt} + j(\omega - \omega_r)\vec{\lambda}_r \qquad (7.29)$$

This means that, in the rotating reference frame, the dq components of the rotor flux $\vec{\lambda}_r$ can be controlled directly by the corresponding dq components of the rotor voltage vector $\vec{v}_r$, after compensating the cross-coupling terms in equation (7.29).

The rotor flux control loops are represented in Figure 7.10. In the q-axis, the set-point value is maintained at $\lambda_{qr}^* = 0$ to align $\vec{\lambda}_r$ with the d-axis, and by comparing it with λ_{qr}, the rotor voltage component v_{qr} is obtained as the sum of the PI regulator output and the cross-coupling term $+s\omega\lambda_{dr}$. In the same way, the component v_{dr} is obtained after calculating the difference between λ_{dr}^*, which is equal to the reference magnitude of the rotor flux, λ_r^*, and λ_{dr}. Being, in this case, the cross-coupling term added to the regulator output $-s\omega\lambda_{qr}$. Hence, using both regulators, and adding the cross-coupling terms, the dq components of the rotor voltage vector $\vec{v}_r$ are obtained.

By applying the inverse Park transformation to such vector, using the slip angle θ_{slip}, the reference phase rotor voltages v_{ar}, v_{br}, v_{cr} are calculated. Then, using pulse width modulation (PWM), the trigger signals $S_{1...6}$ for the RSC switches are obtained. The slip angle is calculated as the difference between the control angle, θ, and the rotor angle, θ_r ($\theta_{slip} = \theta - \theta_r$).

According to the second equation of (7.16), reactive power increases with λ_r, so the DFIG rotor flux magnitude λ_r^* can be obtained from a reactive power controller using a PI regulator, as shown in Figure 7.11, or directly from a voltage controller, as voltage increases with reactive power injection. The same conclusion can be reached using the second equation of equations (7.17) and (7.9): reactive power increases

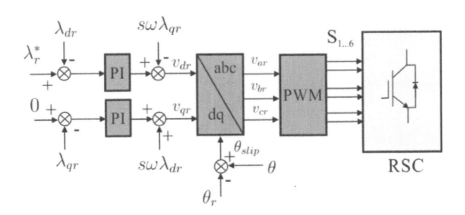

FIGURE 7.10 Rotor flux control loops.

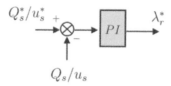

FIGURE 7.11   Reactive power and voltage controller.

by increasing the DFIG internal e.m.f., which is directly proportional to rotor flux. Moreover, in islanding operation, the internal e.m.f. is imposed by the rotor flux magnitude, as given in equation (7.9), and the terminal voltage is obtained through equation (7.10). In this case, the voltage controller will provide the rotor flux magnitude required to compensate the voltage drop in the DFIG equivalent impedance.

7.8   SIMULATION RESULTS

The assessment of the proposed VSM-based control of the DFIG has been done through simulation. A comprehensive model of the DFIG, the proposed control system, and a simulation benchmark has been employed. Figure 7.12 denotes the scheme of the simulated system. This scheme represents a simulation benchmark following the specifications of the Spanish normative for the assessment of the requirements for generators. The simulation benchmark consists of an SG, including its governor and AVR, a load and a transmission line. The DFIG and benchmark parameters are given in the Appendix (Tables 7.1, 7.2 and 7.3, respectively). Note that according to these values, the DFIG is operating in a stiff grid which it is considered challenging for the operation of VSMs.

Simulation results presented below include the response of the DFIG to a wind velocity step at partial load, the response to a step command in the reactive power,

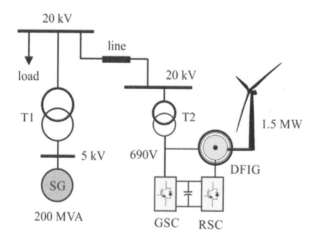

FIGURE 7.12   Scheme of the simulation benchmark.

the response to a change in the grid load, and, finally, the response to the transition from grid-connected to isolated operation. Simulation results validate the proposed rotor flux orientation-based control system for the DFIG. These simulations have been performed using the PSIM software.

7.8.1 DFIG Response to a Step in Wind Velocity at Partial Load

The first test consists of a wind velocity step at partial load. At $t=1$ s, wind velocity changes from 5 to 9 m/s. The wind turbine is initially operating at the maximum power point for 5 m/s (2.1 kNm and 1,083 rpm, at the generator side). Then, because of the wind gust, the wind turbine torque increases, producing an acceleration of the rotor. The increasing rotational speed produces increasing DFIG torque commands following the MPPT strategy. Finally, a new steady-state point is reached at the 9 m/s corresponding maximum power point (6.9 kNm and 1,950 rpm, at the generator side). All the above is illustrated in Figure 7.13.

Moreover, regarding the performance of the proposed control system, this simulation test proves that the control system can maintain the DFIG synchronism under changes in the torque command. As shown in Figure 7.1, the torque command is produced at the MPPT block, and it is the input to the synchronizing block that produces the reference angle θ. Later, the flux control loops maintain the rotor flux oriented to the reference axis and control the rotor flux magnitude. Figure 7.14 denotes that both objectives are met despite the changes in the DFIG torque. The flux orientation to the reference angle θ allows to reach the torque command while maintaining the synchronism.

Finally, Figure 7.15 denotes the rotor currents at the transition around the synchronous speed of the DFIG. The figure clearly illustrates the change in the rotor current phase sequence when changing from sub-synchronous to super-synchronous rotational speed, i.e., from positive to negative slip.

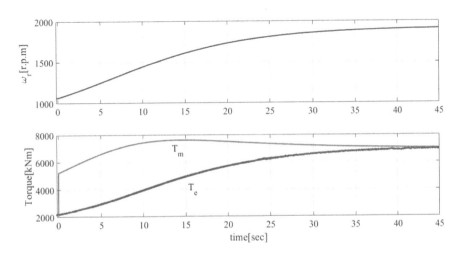

FIGURE 7.13   Wind turbine and DFIG rotational speed and torque.

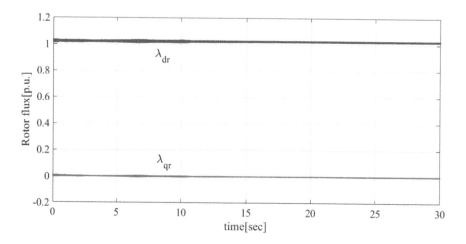

FIGURE 7.14   Rotor flux components.

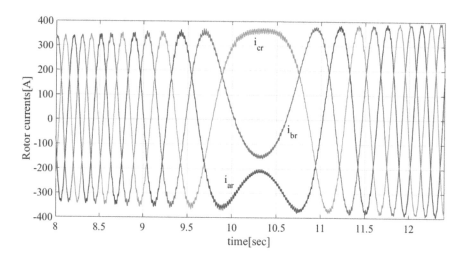

FIGURE 7.15   DFIG rotor currents.

7.8.2   DFIG RESPONSE TO A STEP COMMAND IN REACTIVE POWER

In this subsection, the DFIG response to a reactive power step command is obtained
when the rotor flux magnitude is fixed by a reactive power controller as the one
depicted in Figure 7.11. Initially, the reactive power command is set at 0 MVAr and
at $t = 0.3$ s, a 1,000 kVAr command is set. The time response of the active and reactive
powers is given in Figure 7.16. The wind turbine is operating at a wind velocity of
10 m/s, and the power shown in the figure is the active power of the DFIG stator. The
reactive power reaches the commanded value in a rapid and well-damped manner,

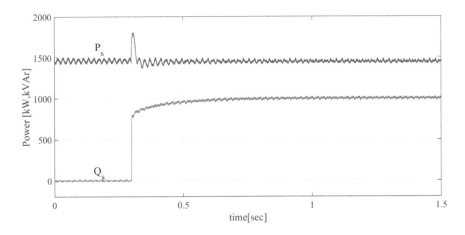

FIGURE 7.16 Active and reactive power response to a reactive power step.

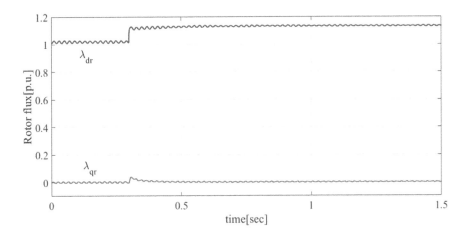

FIGURE 7.17 Rotor flux components response to a reactive power step.

while the active power suffers a slight disturbance due to transient misalignment produced while increasing the rotor flux through the voltage applied by the RSC.

Figure 7.17 denotes the rotor flux components. The direct component increases as demanded by the reactive power controller, while during the transient, a slight misalignment can be observed in the quadrature rotor flux component, producing the active power disturbance noted before.

Finally, Figure 7.18 denotes the stator voltage and current in phase "a". Initially, voltage and current pulse are in phase, because the commanded reactive power is zero. Then, at $t = 0.3$ s, when the reactive power command is changed, the current wave immediately changes to lag the voltage wave as required by the commanded reactive power.

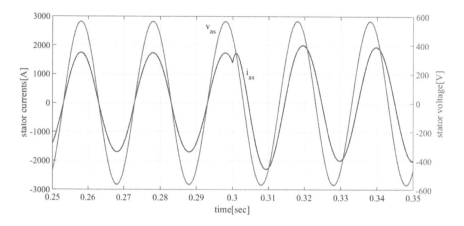

FIGURE 7.18   Detail of instantaneous voltage and current in phase a.

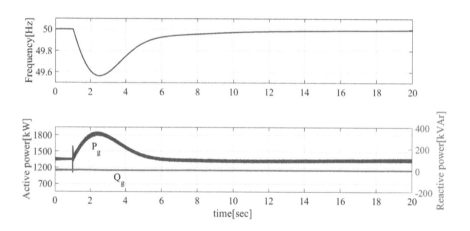

FIGURE 7.19   System frequency and DFIG active and reactive power following a load step.

7.8.3   DFIG Response to a System Load Step

The DFIG response to a system load increase is shown in this subsection. The main purpose of this simulation is to demonstrate that the proposed control system behaves as a voltage source. So, in opposition to the classical vector-oriented control, where the DFIG behaves as a current source supplying constant active and reactive powers, here the generator can "see" the changes in the power system loading. So, at $t = 1$ s, the system load is increased by 0.05 pu. The wind turbine is operating at 9 m/s wind velocity supplying 7 kNm at 1,920 rpm. After the load step, the system frequency drops as depicted in Figure 7.19. Also, in this figure, the DFIG active power increases following the load increment, demonstrating that the DFIG behaves as a voltage source.

Obviously, the DFIG supplies the load increment only temporally because the incoming power of the wind turbine does not change. In fact, the generated power increment is obtained from the kinetic energy of the rotor, producing the deceleration of the rotor, as depicted in Figure 7.20, due to DFIG torque increment, while the WT torque is barely affected, as shown in the figure. The reduction of the rotational speed in turn produces a reduction of the torque reference given by MPPT block of Figure 7.1. This allows the rotor speed recovering, reaching the previous steady state and assuring the wind turbine stability.

Finally, Figure 7.21 denotes the rotor flux components. The response shown in the figure demonstrates the ability of the control system to maintain the rotor flux orientation and magnitude during the DFIG response to a system load step, which in turn is the cause of the DFIG behaving as a voltage source.

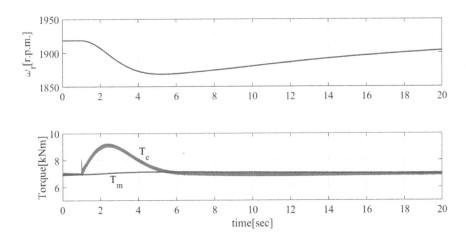

FIGURE 7.20   DFIG rotational speed and torque and wind turbine torque following a load step.

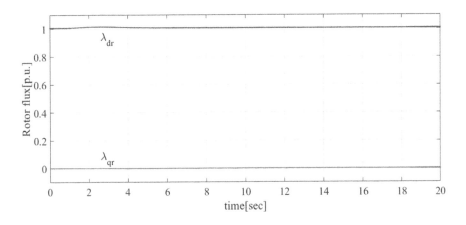

FIGURE 7.21   Rotor flux components following a load step.

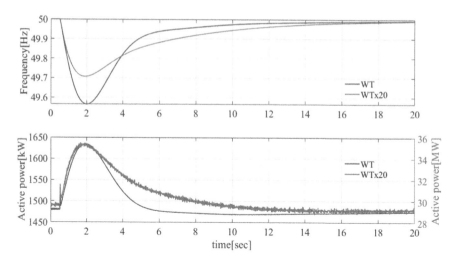

FIGURE 7.22   System frequency and WF active after a load step.

7.8.4   WIND FARM (WF) RESPONSE TO A LOAD STEP

In this subsection, the system frequency response is obtained when a WF, made of 20 WTs, is connected to the system of Figure 7.12, instead of a single wind turbine. A single WT, rated 20×1.5 MW, represents the aggregate model of the WF. Wind velocity, system load, and load increment are the same as those used in Subsection 7.8.3. The objective of this test is to prove the contribution of the proposed control system to the frequency stability when a high proportion of the load is supplied by a WF. For comparison purposes, the system frequency obtained in Section 7.8.3 is also represented here. Thus, Figure 7.22 shows in blue the system frequency when only one wind turbine was connected to the system and in red when the WF is connected. As shown in the figure, frequency response improves in both ROCOF and frequency nadir. The reason for this is that the frequency response of the DFIG is much faster than that of the SG, because of the inherent response delay of the SG governor. This is what in the literature is called fast frequency response. As in subsection 7.8.3, the WF responses to the load increment by increasing its output, alleviating in this way the frequency deviation of the SG.

7.8.5   DFIG RESPONSE TO THE TRANSITION FROM GRID CONNECTED TO ISOLATED

In this subsection, the DFIG will supply the load in isolated mode after the sudden disconnection of the SG. Obviously, the DFIG can supply the load only if the wind velocity is high enough for obtaining the load power. It has to be noted that the DFIG switches from grid-connected to isolated mode automatically, i.e., it is not needed detecting the isolated operation. After the transition, the DFIG change automatically its output, due to its voltage source feature, until the power demanded by the load is met with a certain frequency increment.

Initially, the DIFG is connected to the grid supplying the maximum power, 1,500 kW, for the actual wind velocity, 10 m/s, at 1,938 rpm. Also, the reactive power command is set at 400 kVAr (see Figures 7.23 and 7.27). At $t = 1$ s, the SG is disconnected and the DFIG supplies the load in isolated mode. In isolated operation, the DFIG has to supply exactly the active and reactive powers demanded by the load, including the transformer and line losses. In this case, a resistive 1,000 kW load has been used, while the reactive power is demanded by the transformer (500 kVAr for such load). Figure 7.23 shows the transition from grid connected to isolated operation. The control system allows for a fast and stable response. The synchronizing loop of Figure 7.4 detects a reduction in the DFIG torque, as a consequence of the reduction of the electrical load (see Figure 7.24).

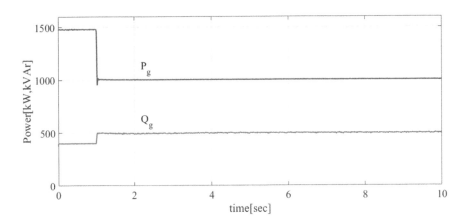

FIGURE 7.23   DFIG active and reactive power.

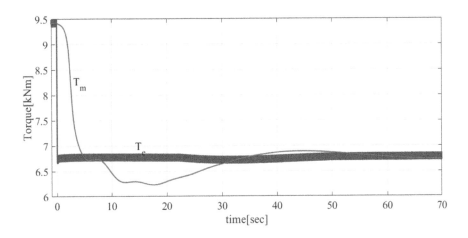

FIGURE 7.24   DFIG and wind turbine torque.

Therefore, the droop control produces a positive frequency increment, as given in Figure 7.25. The reference frequency is increased, and the synchronizing control loop operates with a permanent torque error, as it does not follow the reference torque anymore, but produces the torque required for the current electrical load. On the other hand, the voltage controller of Figure 7.11 automatically obtains the rotor flux magnitude required for supplying the isolated load at the commanded voltage. Figure 7.25 shows the voltage transition, while Figure 7.26 shows the rotor flux components. The q-axis component is kept as its reference value of zero, so the reference frequency is achieved. While the d-axis component produces the desired internal e.m.f. to obtain the reference voltage at the DFIG terminals. Also, it has to be pointed out that if the operation at nominal frequency was desired, a secondary frequency

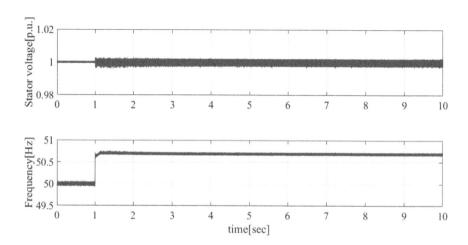

FIGURE 7.25   Voltage and frequency at DFIG terminals.

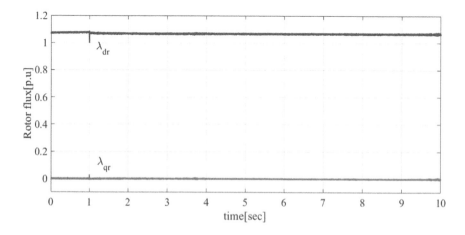

FIGURE 7.26   Rotor flux components.

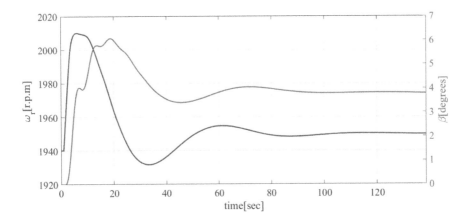

FIGURE 7.27 Wind turbine rotational speed and pitch angle.

controller could have been implemented that would modify the reference frequency ω_0 in Figure 7.4, so that the nominal frequency is obtained after adding the frequency increment $\Delta\omega$.

Moreover, Figure 7.27 shows the wind turbine rotational speed, at the DFIG side, and the pitch angle, while Figure 7.24 shows the wind turbine torque, also at the high-speed shaft. Here, after the transition, the wind turbine accelerates as a consequence of the reduction of the DFIG torque. When the wind turbine reaches its maximum rotational speed, the speed control based on the pitch regulator starts increasing the pitch angle, which reduces the wind turbine driving torque to maintain constant rotational speed. By maintaining constant rotational speed, it can be assured that the power delivered by the wind turbine equals the power demanded by the DFIG, which in turn is equal to the power demanded by the electrical load [11].

These results demonstrate the ability of the proposed control system to switch from grid connected to isolated operation, without having to detect the islanded mode and without changing the control mode. Therefore, these results also prove the operation of the DFIG as a true voltage source using the proposed control system. Finally, to support even more this statement, Figure 7.28 shows the detail of the instantaneous voltage and current during the transition.

7.9 CONCLUSIONS

A novel GFM control strategy for a DFIG based on the orientation of the rotor flux has been proposed in this chapter. The reference axis angle is obtained through the emulation of the SG swing equation, and a flux controller is used to align the rotor flux to the reference axis through the rotor voltage applied by the RSC of the DFIG. By aligning the rotor flux to the reference axis, two control objectives are met: controlling the synchronism of the DFIG and controlling the DFIG torque. Moreover, the flux controller also allows to modify the rotor flux magnitude, which in turn allows to control the DFIG reactive power.

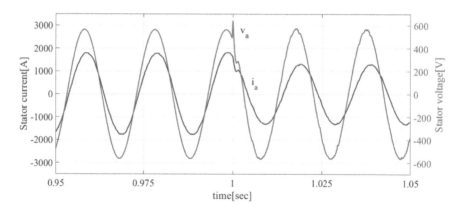

FIGURE 7.28 Detail of instantaneous voltage and current in phase a.

A dynamic model of the DFIG is first presented in the chapter that, in steady state, allows to obtain an equivalent circuit analog to that of the SG. Based on the dynamic model, the DFIG synchronizing torque is obtained, which is used in the synchronization control loop that emulates the swing equation of the SG.

This loop is then linearized and its stability is analyzed. The step response of the linearized system is then obtained, demonstrating the ability of the proposed control system to follow changes in the torque reference and grid frequency disturbances. In fact, this second test also demonstrates the GFM capability of the proposed control scheme because the DFIG modifies its output power following an increment of the grid load, in opposition to the classical vector-oriented control, where the DFIG maintains a constant output independently of the grid load.

Furthermore, the chapter introduces later the inner flux control loops. The rotor flux orientation is obtained through a control loop based on achieving a zero q-axis rotor flux component, i.e., when the q-axis component of the rotor flux is zero, the rotor flux is aligned along the reference axis. However, the d-component of the rotor flux is also controlled, so when the rotor flux is oriented along the reference axis, the d-axis component is equal to the rotor flux magnitude. The rotor voltage components applied by the RSC are used to control the rotor flux components, based on the dynamic equations of the rotor circuit. Also based on the proposed model of the DFIG, the rotor flux magnitude is used to control the DFIG reactive power.

Finally, comprehensive simulation models, including the wind turbine and a benchmark grid, have been used to obtain the time response of the proposed system. The benchmark is based on the Spanish normative for the assessment of the requirements for generators. Simulation results demonstrate the capability of the proposed control system to follow the active and reactive power commands, being the active power commands a consequence of the wind speed variations following the MPPT, and, mainly, to demonstrate its GFM capability, acting as a voltage source under changes in the grid load. Finally, the transition from grid-connected to isolated operation has also been tested to prove the capability of the proposed system to act as a voltage source even in isolated operation. Simulation results also prove the excellent dynamic response of the proposed control system.

APPENDIX

TABLE 7.1
DFIG Parameters

Parameter	Symbol	Value	Unit
Rated power	P_N	2,000	kW
Rated stator voltage	V_s	690	V
Stator frequency	f_s	50	Hz
Poles pair number	p	2	–
Slip range	s	±0.3	pu
Stator resistance	R_s	2.6	mΩ (0.0109 pu)
Rotor resistance (referred to stator)	R'_r	2.9	mΩ (0.0122 pu)
Magnetization inductance	L_m	2.5	mH (3.3 pu)
Stator leakage inductance	$L_{\sigma s}$	0.087	mH (0.0115 pu)
Rotor leakage inductance (referred to stator)	$L'_{\sigma r}$	0.087	mH (0.0115 pu)
Moment of inertia	J	650	kg/m2

TABLE 7.2
Synchronous Generator Parameters

Parameter	Symbol	Value	Units
Rated power	S_N	200	MVA
Rated line-line voltage	V_N	5	kV
Frequency	f_s	50	Hz
Poles pair number	p	2	–
Stator winding resistance	R_s	0.020	pu
Stator leakage inductance	L_s	0.112	pu
d-axis magnetizing inductance	L_{dm}	1.79	pu
q-axis magnetizing inductance	L_{qm}	1.60	pu
Field winding resistance (referred to stator)	R_f	1.21	pu
Field winding leakage inductance (referred to stator)	L_{fl}	0.117	pu
Rotor damping cage d-axis resistance (referred to stator)	R_{dr}	0.03	pu
Rotor damping cage d-axis leakage inductance (referred to stator)	L_{drl}	0.375	pu
Rotor damping cage q-axis resistance (referred to stator)	R_{qr}	0.004	pu
Rotor damping cage d-axis leakage inductance (referred to stator)	L_{qrl}	0.200	pu
Inertia constant	H	2.5	MWs/MVA

TABLE 7.3

Line and Transformer Parameters

Parameter	Symbol	Value	Unit
T1: Rated power	S_N	220	MVA
T1: Rated line-line voltage primary	V_{1N}	20	kV
T1: Rated line-line voltage secondary	V_{2N}	5	kV
T1: Short-circuit ratio	Z_{sc}	10	%
T1: X/R ratio	X/R	30	pu
T1: Frequency	f	50	Hz
T2: Rated power	S_N	2.5	MVA
T2: Rated line-line voltage primary	V_{1N}	20	kV
T2: Rated line-line voltage secondary	V_{2N}	0.26	kV
T2: Short-circuit ratio	Z_{sc}	8	%
T2: X/R ratio	X/R	25	pu
T2: Frequency	f	50	Hz
Line inductance	L_l	3.4	mH
Line resistance	R_l	0.5	Ω

REFERENCES

[1] Pena, R., J. C. Clare, and G. M. Asher. "Doubly fed induction generator using back-to-back PWM converters and its application to variable-speed wind-energy generation." *IEE Proceedings-Electric Power Applications* 143.3(1996): 231–241.

[2] Arnalte, S., J. C. Burgos, and J. L. Rodriguez-Amenedo. "Direct torque control of a doubly-fed induction generator for variable speed wind turbines." *Electric Power Components and Systems* 30.2(2002): 199–216.

[3] Santos-Martin, D., J. L. Rodríguez-Amenedo, and S. Arnaltes. "Dynamic programming power control for doubly fed induction generators." *IEEE Transactions on Power Electronics* 23.5(2008): 2337–2345.

[4] Kroposki, B., et al. "Achieving a 100% renewable grid: Operating electric power systems with extremely high levels of variable renewable energy." *IEEE Power and Energy Magazine* 15.2(2017): 61–73.

[5] Lasseter, R. H., Z. Chen, and D. Pattabiraman. "Grid-forming inverters: A critical asset for the power grid." *IEEE Journal of Emerging and Selected Topics in Power Electronics* 8.2(2019): 925–935.

[6] Wang, S., et al. "On inertial dynamics of virtual-synchronous-controlled DFIG-based wind turbines." *IEEE Transactions on Energy Conversion* 30.4(2015): 1691–1702.

[7] He, Wei, Xiaoming Yuan, and Jiabing Hu. "Inertia provision and estimation of PLL-based DFIG wind turbines." *IEEE Transactions on Power Systems* 32.1(2016): 510–521.

[8] Wang, Shuo, Jiabing Hu, and Xiaoming Yuan. "Virtual synchronous control for grid-connected DFIG-based wind turbines." *IEEE Journal of Emerging and Selected Topics in Power Electronics* 3.4(2015): 932–944.

[9] Nian, Heng, and Yingzong Jiao. "Improved virtual synchronous generator control of DFIG to ride-through symmetrical voltage fault." *IEEE Transactions on Energy Conversion* 35.2(2019): 672–683.

[10] Abad, Gonzalo, et al. *Doubly Fed Induction Machine: Modeling and Control for Wind Energy Generation* (Vol. 85). Hoboken, NJ: John Wiley & Sons (2011).

[11] Arnaltes, Santiago, Jose Luis Rodriguez-Amenedo, and Miguel E. Montilla-DJesus. "Control of variable speed wind turbines with doubly fed asynchronous generators for stand-alone applications." *Energies* 11.1(2018): 1–26.

8 Damping Power System Oscillations Using Grid-Forming Converters

José Luis Rodríguez Amenedo
and Santiago Arnaltes
University Carlos III of Madrid

CONTENTS

8.1 INTRODUCTION

Nowadays, it is a well-known fact that the high penetration of renewable power plants has completely changed the electrical generation model. In some power systems, renewable energy sources (RES), based on power electronic converters (PECs), are displacing conventional synchronous generators (SGs) reducing system inertia and fault current contribution [1]. Although RES are increasing their contribution to grid ancillary services, in order to maintain a safe and reliable power systems operation, some services, such as power system restoration or low-frequency oscillation (LFO) damping, are requested only to SGs. Nevertheless, new grid-forming converters (GFCs), emulating SGs behavior, can contribute to provide these services as well [2–8].

This chapter proposes two power system stabilizers (PSSs) for damping LFO in power systems using PECs. First, a power stabilizer, implemented in a grid-forming converter, has been used to provide LFO damping by acting on the converter power

DOI: 10.1201/9781003302520-8

angle and exchanging active power (POD-P). On the other hand, a second power stabilizer provides LFO damping through the reactive power (POD-Q) when the power converter is used as a STATCOM using the voltage control loop to damp power oscillations. Both power stabilizers have been compared in two case studies. In the first case, LFO damping in a single machine connected to an infinite bus through a tie-line is analyzed, and in the second case, damping of inter-area oscillation modes, in a two-area system, is also studied. Simulation results and stability studies demonstrate that both stabilizers damp power oscillations, being the POD-P stabilizer more effective than the POD-Q, although the former requires some kind of energy provision in the DC bus. System description and modeling of an SG and a PEC, both connected in parallel to the same bus, are shown in the following section. Next, control schemes of the grid-forming POD-P and STATCOM POD-Q stabilizers are proposed. From these two stabilizers, a small-signal analysis is carried out, studying the influence of both stabilizers in the generator's electromechanical and flux-decay modes. This section ends with a time-domain simulation comparing the damping of LFO modes in each case.

Subsequently, the damping performance of both stabilizers is checked in a two-area system. Both areas are interconnected through a transmission line. The dynamic response of this system is analyzed under a load change in one area and during a line tripping. As in the previous case, an eigenvalues analysis is also carried out when the reactance of the interconnection line changes, with and without POD compensation.

8.2 SYSTEM DESCRIPTION AND MODELING

Figure 8.1 shows the connection of an SG and a PEC to an infinite bus through a reactance X_e with the purpose of damping power oscillations. Both devices are connected in parallel to the same bus where the voltage vector is represented by $V_t e^{j\theta}$, being V_t the voltage magnitude and θ its angle with respect to infinite bus $V_\infty \angle 0°$. The current vector through the reactance X_e is $I_G e^{j\gamma}$.

In LFO studies, it is common to use SGs reduced-order models. Here, a simple one-axis (flux-decay) model has been considered. This is a four-order model [9] using E'_q as a state variable that represents the field flux dynamics but oriented to the quadrature axis, q. The position of this q-axis with respect to the reference voltage $V_\infty \angle 0°$ is the angle δ that represents the second variable state. The dynamic

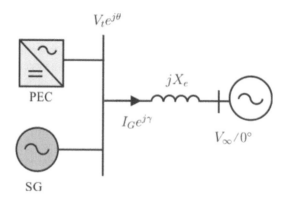

FIGURE 8.1 SG and PEC in a single-machine infinite-bus system.

model is completed by defining the generator speed ω and the excitation voltage E_{fd} as the other state variables. In this model, faster dynamics, including stator/network dynamics, have been neglected. From a complete set of nonlinear SG dynamic equations, a state-space model is obtained by linearizing them at a given operating point. The state matrix eigenvalues determine the oscillation modes that can be damped using suitable power stabilizers. The dynamic equations of the model are shown below.

The dynamic equation of E_q' can be expressed as

$$T'_{d0}\frac{dE_q'}{dt} = -E_q' - \left(X_d - X_d'\right)I_d + E_{fd} \tag{8.1}$$

where T_{d0}' is the d-axis open-circuit transient time constant. The reactance X_d and X_d' are the total and transient d-axis reactance, respectively. The equation of motion of an SG is as follows:

$$2H\frac{d\bar{\omega}}{dt} = T_m - T_e - D\left(\bar{\omega} - \bar{\omega}_0\right) \tag{8.2}$$

where H is the inertia constant of the generator, expressed in seconds, D is the damping constant, and $\bar{\omega}$ is the angular velocity of the rotor, and $\bar{\omega}_0$ is the reference angular speed, both magnitudes expressed in pu. Meanwhile, T_m is an input of the model and represents the mechanical torque also in pu. T_e is the electrical torque, which is expressed as

$$T_e = E_q'I_q + \left(X_q - X_d'\right)I_d I_q \tag{8.3}$$

Note that the electrical torque is calculated as a nonlinear expression of electrical variables and the SG parameters. X_q is the total q-axis reactance, and I_d and I_q are the dq-components of the current vector. The dynamic equation of δ is obtained by integrating the rotational speed of the generator, ω, with respect to a synchronously rotating reference, ω_s, in rad/s. In pu, the dynamic equation of δ is as follows

$$\frac{1}{\omega_s}\frac{d\delta}{dt} = \bar{\omega} - \bar{\omega}_0 \tag{8.4}$$

where $\bar{\omega}$ is the speed expressed in pu and $\bar{\omega}_0 = 1$ pu. Assuming a fast exciter for regulating voltage V_t at the output terminal from the reference voltage V_{ref}, the dynamic equation of E_{fd} is equal to

$$T_A\frac{dE_{fd}}{dt} = -E_{fd} + K_A\left(V_{ref} - V_t\right) \tag{8.5}$$

where T_A and K_A are the time constant and the gain of the exciter.

Figure 8.2 shows the vector diagram of the voltage and current of the SG. The q-axis of the generator forms an angle δ with respect to a D-axis (aligned to the

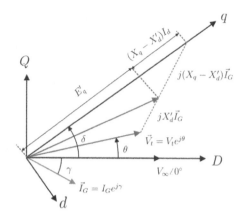

FIGURE 8.2   SG voltage and current diagram.

reference voltage vector $V_\infty \angle 0°$ at the infinite bus). The angle θ corresponds to the voltage $\vec{V}_t = V_t e^{j\theta}$ at the generator terminals.

Next expressions complete the set of algebraic equations

$$X_q I_q - V_d = 0 \tag{8.6}$$

$$E'_q - V_q - X'_d I_d = 0 \tag{8.7}$$

Figure 8.3 shows an equivalent circuit of the SG and the PEC corresponding to the single-machine infinite-bus system of Figure 8.1. The SG is represented as a dependent voltage source $\vec{E}_{SG}$ [9]

$$\vec{E}_{SG} = \left[\left(X_q - X'_d \right) I_q + j E'_q \right] e^{j\left(\delta - \frac{\pi}{2} \right)} \tag{8.8}$$

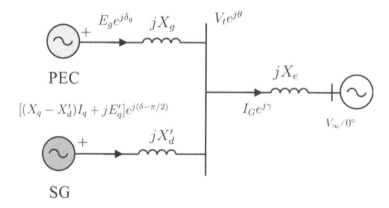

FIGURE 8.3   Equivalent circuit of a one-axis SG and a PEC.

behind the reactance X'_d. Similarly, the PEC is modeled as an independent voltage source $E_g e^{j\delta_g}$ behind a reactance X_g.

The voltage vector $V_t e^{j\theta}$, shown in Figure 8.3, is obtained by applying Millman's theorem in terms of $E_g e^{j\delta_g}$ and $V_\infty \angle 0°$ as follows:

$$V_t e^{j\theta} = \left(V_d + jV_q\right)e^{j\left(\delta-\frac{\pi}{2}\right)} = +K'_d\left[\left(X_q - X'_d\right)I_q + jE'_q\right]e^{j\left(\delta-\frac{\pi}{2}\right)}$$

$$+ K_g jE_g e^{j\left(\delta_g-\frac{\pi}{2}\right)} + K_e V_\infty \qquad (8.9)$$

where K'_d, K_g, and K_e are

$$K'_d = \left(\frac{\overline{Y}'_d}{\overline{Y}'_d + \overline{Y}_g + \overline{Y}_e}\right) = \frac{X_g X_e}{X_g X_e + X'_d X_e + X'_d X_g} \qquad (8.10)$$

$$K_g = \left(\frac{\overline{Y}_g}{\overline{Y}'_d + \overline{Y}_g + \overline{Y}_e}\right) = \frac{X'_d X_e}{X_g X_e + X'_d X_e + X'_d X_g} \qquad (8.11)$$

$$K_e = \left(\frac{\overline{Y}_e}{\overline{Y}'_d + \overline{Y}_g + \overline{Y}_e}\right) = \frac{X'_d X_g}{X_g X_e + X'_d X_e + X'_d X_g} \qquad (8.12)$$

being the admittances $\overline{Y}'_d = -j/X'_d$, $\overline{Y}_g = -j/X_g$ and $\overline{Y}_e = -j/X_e$. As these admittances have only imaginary part K'_d, K_g and K_e appear as constants in equation (8.9).

By multiplying equation (8.9) by $e^{-j\left(\delta-\frac{\pi}{2}\right)}$ and separating into real and imaginary parts, the following algebraic equations of the network are obtained:

$$V_d - K'_d\left(X_q - X'_d\right)I_q + K_g E_g \sin\left(\delta_g - \delta\right) - K_e V_\infty \sin\delta = 0 \qquad (8.13)$$

$$V_q - K'_d E'_q - K_g E_g \cos\left(\delta_g - \delta\right) - K_e V_\infty \cos\delta = 0 \qquad (8.14)$$

PEC's active and reactive power expressions, expressed in pu, are

$$P_g = \left(\frac{V_t}{X_g}\right)E_g \sin\left(\delta_g - \theta\right) \qquad (8.15)$$

$$Q_g = \left(\frac{V_t}{X_g}\right)\left[E_g \cos\left(\delta_g - \theta\right) - V_t\right] \qquad (8.16)$$

According to equations (8.15) and (8.16), it is well known that the active power P_g is controlled acting on the angle δ_g and the reactive power Q_g by the voltage control of E_g.

PEC control is designed to damp power oscillations by exchanging active and reactive power, acting on the angle and module of its internal voltage. In this chapter, two stabilizers are compared, one implemented in a grid-forming converter and the other in a STATCOM. The fundamentals of both stabilizers are presented below.

8.3 GRID-FORMING CONVERTER STABILIZER

Unlike conventional grid-following (GFL) converters, grid-forming converters operate as a real voltage source maintaining an internal voltage phasor that enables to immediate response to changes in the external system without using any phase-locked loop (PLL) to synchronize it to the grid. Instead of a PLL, a grid-forming converter uses a synchronization loop based on the active power response as SGs do.

Figure 8.4 shows the synchronization loop where the power angle δ_g is obtained from the difference of active power reference P_g^* and the actual power P_g. As previously mentioned, δ_g defines the angular position of the internal voltage vector $E_g e^{j\delta_g}$ with respect to reference voltage $V_\infty \angle 0°$. According to equation (8.15), the active power P_g depends on δ_g and the voltage vector $V_t e^{j\theta}$. The derivative of angle δ_g is obtained as the difference between the internal angular speed ω_g and the angular speed corresponding to the reference frequency ω_s.

Considering that $\bar{\omega}_g$ is expressed in pu and $\bar{\omega}_0 = 1$ pu, the power angle δ_g is calculated as

$$\frac{1}{\omega_s}\frac{d\delta_g}{dt} = \bar{\omega}_g - \bar{\omega}_0 \tag{8.17}$$

This equation is completely analogous to equation (8.4) that defined the SG.

As shown in Figure 8.4, the internal frequency $\bar{\omega}_g$ is obtained in terms of P_g^* and P_g as

$$\bar{\omega}_g = \bar{\omega}_0 + \frac{P_g^* - P_g}{J_g s + D_g} \tag{8.18}$$

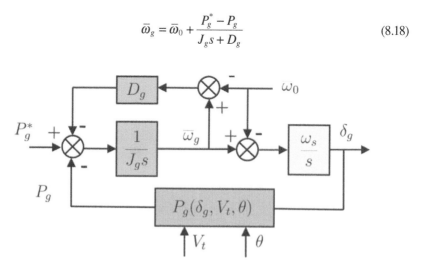

FIGURE 8.4 Power synchronization loop of a grid-forming converter.

where J_g is the inertia moment of the grid-forming (in seconds), and D_g is denoted as a damping constant. The inverse of the damping constant $R_g = 1/D_g$ is known as the droop constant that, in pu, is the ratio of frequency variation in pu, $\Delta f / f_n$, and the normalized power deviation $\Delta P / P_n$. When the inertia constant is equal to zero, $\bar{\omega}_g$ is calculated as the product of a droop constant and the active power increment with respect to a reference, $\bar{\omega}_g - \bar{\omega}_0 = R_g \left(P_g^* - P_g \right)$. As shown in Figure 8.5, in a grid-forming converter, LFO damping is achieved by acting on the internal frequency reference of the power synchronization loop. The signal $\bar{\omega}'$, that is subtracted to the internal frequency $\bar{\omega}_g$, is obtained from the measured frequency ω at the output terminals of the grid-forming converter through a washout function as follows:

$$\bar{\omega}' = K_w \left(\frac{T_w s}{T_w s + 1} \right) \bar{\omega} \tag{8.19}$$

In an SG, its PSS uses a function similar to equation (8.19), but here a phase compensation block is not needed, because in an SG, a lead-lag function is needed to compensate the delay of the exciter effect on the generator torque. However, in a grid-forming converter, it is not required since an immediate action on the active power is obtained when the internal frequency $\bar{\omega}_g$ is modified. The derivative term in the washout filter expressed in equation (8.19) blocks any DC component existing

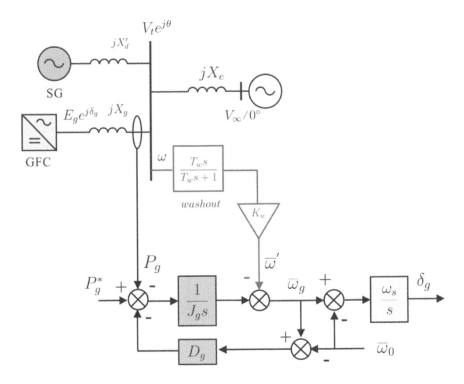

FIGURE 8.5 Block diagram of a grid-forming stabilizer.

on $\bar{\omega}$ as a consequence of a permanent deviation from the rated frequency ω_s. Furthermore, the time constant T_w is chosen in a range of 0.1–20 s so that LFOs measured in the bus can pass unaltered. So, this stabilizer operates only in the range of low-frequency oscillations, while the gain K_w is tuned in order to provide enough damping to the oscillation modes.

The proposed stabilizer acts directly over δ_g and therefore on the active power. Note that a pole-zero cancelation occurs between the stabilizer filter and the integral function ω_s/s. In a GFL converter, a stabilizer of this type can be implemented by calculating the power reference as a function of the frequency measured on the bus. However, its effectiveness is lower due to the dynamics between P_g^* and P_g. As mentioned above, here, the stabilizer acts on angle δ_g, keeping voltage magnitude E_g constant, and also if $E_g = V_t$, according to equation (8.16), the reactive power Q_g is approximately zero, since the angular difference between δ_g and θ is usually small. In the linearized grid-forming model, in order to simplify the study, a zero-inertia has been considered.

Considering this assumption, the frequency deviation $\Delta\omega_g$ is equal to

$$\Delta\omega_g = -R_g\Delta P_g = -R_g K_s\left(\Delta\delta_g - \Delta\theta\right) \tag{8.20}$$

where ΔP_g is obtained by linearizing equation (8.15) when E_g is kept constant, being the synchronization constant K_s

$$K_s = \left(\frac{V_t^0 E_g^0}{X_g}\right)\cos\left(\delta_g^0 - \theta^0\right) \tag{8.21}$$

Assuming that $X_g = 0.15$ pu, $V_t^0 = E_g^0 = 1$ pu, and $\delta_g^0 = \theta^0$, the synchronization constant value is $K_s = 6.67$ pu. According to the values given in Table 8.3 (Annex A), K_s is significantly higher than the synchronization constant on an SG, which means that the active power transmission is achieved with lower δ_g angles.

Figure 8.6 shows a block diagram of the grid-forming stabilizer represented in equation (8.19). The output from this block is the frequency compensation variable $\Delta\bar{\omega}'$, that is proportional to the difference between $\Delta\bar{\omega}$ and Δz, as

$$\Delta\bar{\omega}' = K_w\left(\Delta\bar{\omega} - \Delta z\right) \tag{8.22}$$

where Δz is an auxiliary variable state, and K_w is the stabilizer gain.

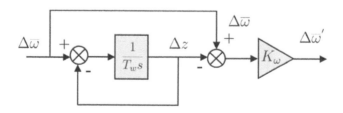

FIGURE 8.6 Block diagram of the washout filter.

A linearized block diagram of the power synchronization loop for a grid-forming converter is shown in Figure 8.7. This figure includes the POD-P active power stabilizer block. Furthermore, it is considered that the linearized active power reference ΔP_g^* is zero, and according to equation (8.20), the actual active power ΔP_g is proportional to the difference of angles $\Delta \delta_g$ and $\Delta \theta$.

By linearizing the expression $V_d = V_t \sin(\delta - \theta)$, the incremental variable $\Delta \theta$ is obtained in terms of $\Delta \delta$ and the algebraic variables ΔV_d and ΔV_q as follows

$$\Delta \theta = \Delta \delta + \left(\frac{E_g^0}{X_g K_s} \right) \left[\Delta V_t \sin\left(\delta^0 - \theta^0 \right) - \Delta V_d \right] \tag{8.23}$$

where ΔV_t is expressed as a function of ΔV_d and ΔV_q according to

$$\Delta V_t = \left(\frac{V_d^0}{V_t^0} \right) \Delta V_d + \left(\frac{V_q^0}{V_t^0} \right) \Delta V_q \tag{8.24}$$

The GFC connection increases the number of dynamic equations by two. Linearizing equation (8.17) and considering equation (8.20), the dynamic equation of $\Delta \delta_g$ is obtained as

$$\frac{1}{\omega_s} \frac{d\Delta \delta_g}{dt} = -R_g K_s \left(\Delta \delta_g - \Delta \theta \right) - K_w \left(\Delta \bar{\omega} - \Delta z \right) \tag{8.25}$$

On the other hand, a second dynamic equation corresponding to the auxiliary variable Δz is expressed as

$$T_w \frac{d\Delta z}{dt} = \Delta \bar{\omega} - \Delta z \tag{8.26}$$

Thus, the complete set of dynamic equations correspond to a sixth-order model, which consists of the grid-forming equations (8.25) and (8.26) and the four SG equations (8.1), (8.2), (8.4), and (8.5). While the algebraic equations of the system, that

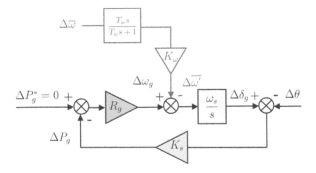

FIGURE 8.7   Linearized block diagram. Grid-forming converter with a POD-P stabilizer.

complete the model, include equations (8.6) and (8.7), which correspond to the SG, and another two that are obtained by linearizing equations (8.13) and (8.14) considering that E_g is constant. Finally, matrices A, B, C, and D of the linearized model are obtained using this set of differential and algebraic equations. The analytical expression of these matrices is given in Appendix B.

8.4   STATCOM STABILIZER

Power system oscillations damping through a STATCOM is carried out acting on the voltage magnitude similarly to conventional PSS of SGs, but without using any phase compensator. In the case of SGs, a PSS modifies the active power generated acting on the voltage E_g (equation 8.15) through the exciter and damping the generator rotor angle swings. A STATCOM can produce the same effect by modifying the bus voltage through the injection of reactive power.

STATCOM models, in transient stability studies, are usually represented by some algebraic equations and the dynamics of the DC voltage. For the sake of simplicity, in this analysis, the dynamic of the DC voltage has not been considered so that the STATCOM only exchanges reactive power in such a way that $\delta_g = \theta$. According to equation (8.15), the active power is proportional to $sin(\delta_g - \theta)$ so that this active power is zero when both angles are the same. LFO damping is now achieved by changing the voltage reference E_g^* through the signal E_g' which is the output of a washout filter shown in Figure 8.8.

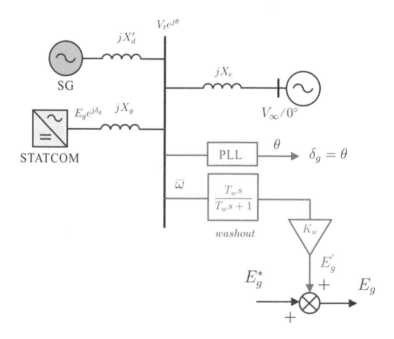

FIGURE 8.8   Block diagram of a STATCOM stabilizer.

Considering that $\delta_g = \theta$ and according to equation (8.16), the reactive power exchange by the STATCOM only depends on voltage E_g with respect to bus voltage V_t.

$$Q_g = \frac{V_t}{X_g}\left(E_g - V_t\right) \qquad (8.27)$$

From the frequency measured in the bus, $\bar{\omega}$, and using the same washout filter, a compensation voltage E'_g is obtained as follows

$$E'_g = K_w\left(\frac{T_w s}{T_w s + 1}\right)\bar{\omega} \qquad (8.28)$$

As shown in Figure 8.8, the signal E'_g is added to the internal voltage reference E^*_g. Unlike to the grid-forming stabilizer, in this case, the LFO damping is achieved by acting on the voltage instead of the frequency. In this case, the compensator is denoted as POD-Q since the voltage variation leads to reactive power exchange.

Linearizing equation (8.28), $\Delta E'_g$ is now different to zero and equal to

$$\Delta E_g = K_w\left(\Delta\bar{\omega} - \Delta z\right) \qquad (8.29)$$

where Δz is the same auxiliary variable used in equation (8.22).

Now, a fifth-order model completes the set of dynamic equations. Only one additional equation, corresponding to the washout filter, is considered equation (8.26). All dynamic equations of the SG and the four algebraic equations of the previous case are the same. In this case, matrices $A, B, C,$ and D are different since now $\Delta E'_g$ is not null and depends on $\Delta\bar{\omega}$ and Δz. The expression of the matrices of the linearized model is also given in Appendix B.

8.5 STABILITY ANALYSIS

8.5.1 SMALL-SIGNAL STABILITY

The aim of this section is to analyze the small-signal stability of the linearized systems described previously. An eigenvalues analysis will be carried out in order to evaluate the impact on the oscillation modes when using the stabilizers associated to grid-forming converters and STATCOM. All parameters used in the model are shown in Table 8.3 of Annex A.

System eigenvalues are obtained from the state matrix which is calculated using the parameters of Table 8.3 and the state and algebraic variables evaluated at the operation point. Table 8.4 of Annex A gives these state and algebraic variables in an operation point corresponding to a voltage at the generator bus with module $V_t = 1$ pu and angle $\theta = 9^0$.

Figure 8.9 shows, in blue, the eigenvalues of a single-machine infinite-bus case corresponding to the electromechanical and the flux-decay oscillation modes of an SG. These modes are also represented when using the stabilizers POD-P (grid-forming) and POD-Q (STATCOM). In the case of the grid forming, the oscillation modes are represented in red, and for the STATCOM, the oscillation modes are represented in green.

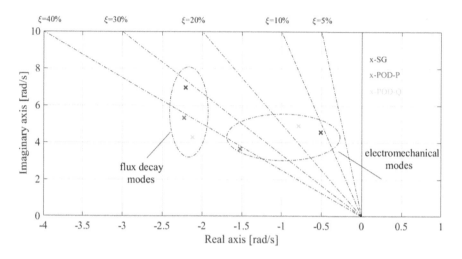

FIGURE 8.9   Eigenvalue loci for a single-machine infinite-bus system.

This figure shows that the electromechanical oscillation mode is significantly damped up to a damping factor of 40.2% when a POD-P is used if compared to the baseline case defined as an SG without any PEC stabilizer. In the case of using a POD-Q with a STATCOM, the electromechanical oscillation mode is only damped up to 15.84%, which demonstrates that it is not so effective as the POD-P stabilizer. However, a POD-Q is very effective to damp the oscillation modes associated to the flux-decay since a POD-Q acts directly on the voltage. The damping factor of this mode is significantly high, 44.47%, in the case of the POD-P.

In Table 8.1 system eigenvalues, damping factor and natural frequency, as well as the corresponding participation factors for each state variable, are shown considering the three cases analyzed. In the baseline case, eigenvalues λ_{12}, corresponding to the electromechanical mode, are dominated by the states $\Delta\bar{\omega}$ and $\Delta\delta$ with a participation factor of 0.38 each one. While for the flux-decay mode λ_{34}, $\Delta E'_q$ and ΔE_{fd} are their dominant states with a participation factor of 0.41 and 0.39, respectively.

The eigenvalue analysis, when a POD-P stabilizer is used, shows that participation factors of electromechanical and flux-decay modes are more distributed between states $\Delta E'_q$, $\Delta\bar{\omega}$, $\Delta\delta$, and ΔE_{fd}. Besides, in this case, two additional non-oscillatory stable eigenvalues, λ_5 and λ_6, are observed.

The eigenvalue λ_5 is dominated by the state $\Delta\delta_g$, and λ_6 by the auxiliary state Δz. In the first case, the participation factor of the dominant state is 0.95 with natural frequency of 6.74 Hz, and for the second one, the participation factor of Δz is 0.96 with a natural frequency of 0.16 Hz. These two eigenvalues have not been represented in Figure 8.9.

A similar study has been carried out for a POD-Q stabilizer. The electromechanical modes are slightly modified by the STATCOM action, and the participation factor of the dominant states is very similar to the baseline case. However, for the flux-decay mode, states $\Delta E'_q$ and ΔE_{fd} are no longer so dominant and a higher

TABLE 8.1

Eigenvalues, Damping Factor and Natural Frequency, as well as the Participation Factors for the Three Cases Analyzed

Eigenvalues	ξ (%)	ω_n (Hz)	$\Delta E'_q$	$\Delta\delta$	$\Delta\bar\omega$	ΔE_{fd}	$\Delta\delta_g$	Δz
			SG (baseline case)					
$\lambda_{12} = -0.51 \pm j4.55$	11.17	0.72	0.14	0.38	0.38	0.10	0.00	0.00
$\lambda_{34} = -2.21 \pm j6.95$	30.25	1.10	0.41	0.10	0.10	0.39	0.00	0.00
GFC (POD-P)								
$\lambda_{12} = -1.52 \pm j3.63$	40.20	0.58	0.32	0.25	0.17	0.10	0.01	0.02
$\lambda_{34} = -2.22 \pm j5.33$	40.30	0.85	0.24	0.27	0.24	0.20	0.03	0.02
$\lambda_5 = -42.38$	100	6.74	0.00	0.00	0.05	0.00	0.95	0.00
$\lambda_6 = -1.03$	100	0.16	0.00	0.04	0.00	0.00	0.00	0.96
			STATCOM (POD-Q)					
$\lambda_{12} = -0.78 \pm j4.90$	15.84	0.78	0.19	0.37	0.31	0.12	0.00	0.01
$\lambda_{34} = -2.12 \pm j4.27$	44.47	0.68	0.39	0.17	0.10	0.32	0.00	0.02
$\lambda_5 = -0.95$	100	0.15	0.00	0.04	0.00	0.00	0.00	0.96

contribution of $\Delta\delta$ is observed. A non-oscillatory mode λ_5, corresponding to the washout filter, appears in this case. Its natural frequency is 0.15 Hz and has not been represented in Figure 8.9 either.

8.5.2 Time-Domain Response

In this section, the time-domain response of active and reactive power transmitted through a line in a single-machine infinite-bus configuration is analyzed. At $t = 0$, a step change of 0.5 pu is applied to the generator torque reference. Figure 8.10 shows the active power response for the three cases considered.

As shown in Figure 8.10, the POD-P stabilizer damps the active power oscillation in just 3 s with only two oscillations before reaching steady state. On the contrary, POD-Q damps the active power oscillation very slightly in comparison to the baseline case. In this case, the steady state is reached in about 10 s.

Figure 8.11 shows a similar response corresponding to the reactive power. The response times and oscillation modes are consistent with the eigenvalue analysis performed in the previous subsections.

8.6 TWO-AREA SYSTEM

In this section, the damping of the inter-area oscillations is addressed by applying the control schemes proposed above. Inter-area oscillations usually appear when two synchronous areas are interconnected through a weak tie-line.

Figure 8.12 shows a single line diagram where two synchronous areas, represented by SG1 and SG2, are interconnected through a double interconnection line.

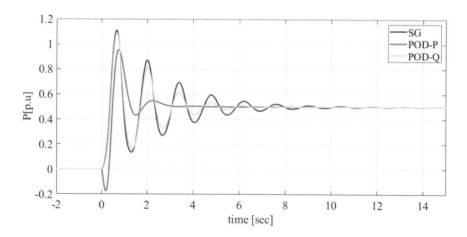

FIGURE 8.10   Active power response to a torque step change of 0.5 pu in the SG.

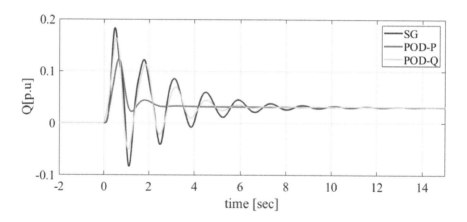

FIGURE 8.11   Reactive power response to a torque step change of 0.5 pu in the SG.

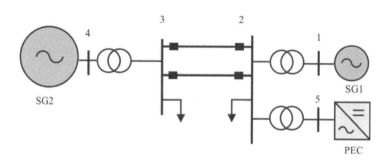

FIGURE 8.12   Single line diagram of two-area system with a PEC.

The equivalent generator SG1 has a power rating of 1,500 MVA, and it is connected to the bus 1 through a step-up transformer. At the bus 1 is also connected a PEC with the same power rating than SG1, 1,500 MVA, using another transformer. Likewise, the equivalent generator SG2 is connected to the other end of the line and has a power rating of 4,000 MVA. In Table 8.2, it is shown the steady-state operation point of the two-area system before applying a load change. Each line between buses 2 and 3 transmits 50 MW, being the active and reactive power injected by the PEC equal to zero.

In this study, a variable reactance, X_L, corresponding to each transmission line, has been considered. Its value varies from 0.01 to 0.9 pu, considering a power rating of 100 MVA and a rated voltage of 400 kV. The step-up transformer of SG2 has a short-circuit impedance of 0.3%, and 1% for the SG1 transformer. In both cases, a 100 MVA power base has been considered. The loads have been modeled as ZI loads, considering constant current when modeling active power changes and constant impedance for reactive power changes.

SG1 and SG2 have been modeled using the well-known GENROU model. The steam turbine and the speed governor have been modeled using the IEEEG1 model. And for the excitation system, a model IEEE-ST1 has been used. In Appendix C, all the parameters of these models are given.

8.6.1 Eigenvalues Loci Under Variation of Parameters

Figure 8.13 shows the oscillation modes of the two-area system in the three cases previously analyzed and when the reactance X_L of the transmission lines varies between 0.01 and 0.9 pu First, the electromechanical and flux-decay oscillation modes of generators SG1 and SG2 are shown when no PEC is connected. Subsequently, these oscillation modes are represented when a PEC is connected to bus 2, first acting as a grid forming using a POD-P stabilizer, or after as STATCOM applying a POD-Q control.

Small-signal stability analysis of power systems can be studied using different methodologies. Among them is worth to mention the frequency-domain methods, based on the impedance model, and phase-amplitude dynamics model, or the time-domain methods based on the state-space model [10]. This last one has been used in this section. Besides, as in the previous sections, the small-signal stability analysis

TABLE 8.2
Steady-State Operation of Generators and Loads

Generators	MW	Vt
BUS1	1350	1.0
BUS2	3900	1.0
BUS3	0.0	1.0
LOADS	MW	MVar
BUS2	1250	0.0
BUS3	4000	0.0

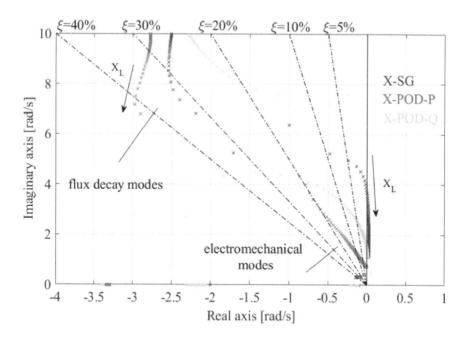

FIGURE 8.13 Eigenvalue loci of the two-area system under X_L variation.

corresponding to the two-area system could be addressed analytically; however, a numerical method has been used in order to avoid making the exposition overly cumbersome.

The eigenvalue loci corresponding to LFO of the two-area system is shown in Figure 8.13. In this figure, the three cases analyzed are represented when X_L varies from 0.01 to 0.9 pu. As shown in Figure 8.13, when the interconnection between areas is weaker (higher X_L value), the natural frequency of the mechanical oscillation modes (inter-area modes) is reduced, as well as the damping factor. If a stabilizer is not applied, the system could become unstable when the line has a reactance higher than 0.13 pu.

When a POD-P stabilizer is used on the PEC, acting as a grid-forming converter, the damping factor corresponding to the electromechanical modes increases up 20% regardless of X_L. As for the flux-decay oscillation modes, the same damping effect is also observed. However, in the case of using a POD-Q, damping of the flux-decay oscillation modes is reduced when X_L increases. In conclusion, POD-P is more effective than POD-Q in all the cases analyzed.

8.6.2 DYNAMIC RESPONSE TO A LOAD CHANGE

In this section, the dynamic response of the two-area system is analyzed when a power reduction of 100 MW is applied to the load connected in bus 2 at $t = 0$. The initial conditions, before the load change, are those shown in Table 8.2. The reactance chosen for the transmission line is $X_L = 0.1$ pu (0.2 pu in each line) since gives rise to a stable operation even in case that no stabilizer is used.

Figure 8.14 shows the active and reactive power transmitted through one of the transmission lines after a load change. The dynamic response of both powers is represented in the three cases analyzed. In the first case, when the PEC is disabled (SG case), power oscillations show a frequency of 0.64 Hz and are weakly damped. In the second case, corresponding to the POD-P stabilizer, LFO is reduced reaching a damping factor of 20%. In case of using a POD-Q stabilizer, the power oscillations are also reduced in comparison to the first case; however, it is less effective than the POD-P. Initially, each interconnection line transmits 50 MW and 5 MVar. After the load change, the power transmitted is 100 MW and 35 MVar in steady state.

Figure 8.15 shows the active and reactive power generated by SG1. Active power remains constant in steady state after the load change (1,350 MW) and the reactive power increases slightly. The reactive power presents poorly damped oscillations

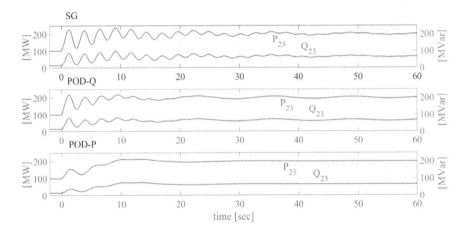

FIGURE 8.14 Active and reactive power transmitted by the two transmission lines under a load change.

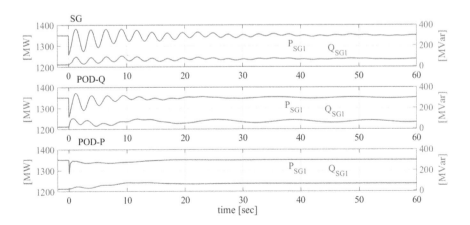

FIGURE 8.15 Active and reactive power response of SG1.

when a POD-Q is used, which is in accordance with the eigenvalue analysis shown in Figure 8.13.

As before, Figure 8.16 shows the active and reactive power generated by SG2. In this case, the active power is reduced by 100 MW (from 3,900 to 3,800 MW) in such a way that the load at node 3 (4,000 MW) is covered with 200 MW from the transmission lines and 3,800 MW from SG2.

Figure 8.17 shows the active and reactive power exchange of the PEC when a POD-P or a POD-Q stabilizer is used. In the first case, the grid-forming converter only exchanges active power, being reactive power equal to zero. After the load change, the GFC absorbs a maximum power of 70 MW and generates a maximum power of 30 MW, and stop exchanging active power when the frequency reaches steady state. In the STATCOM case, its behavior is just the opposite, it only exchanges reactive power being the active power exchange equal to zero. The reactive power exchange varies by ±25 MVar.

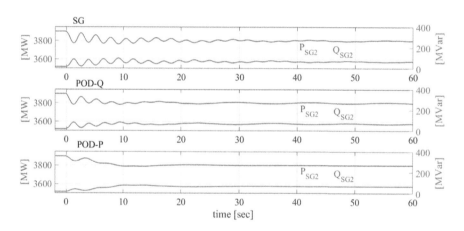

FIGURE 8.16 Active and reactive power response of SG2.

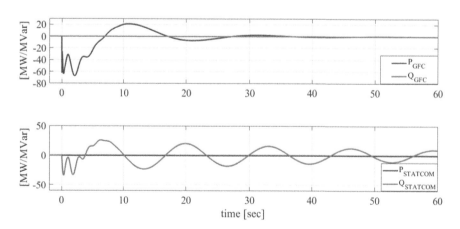

FIGURE 8.17 Active and reactive power exchange by a GFC and a STATCOM.

In conclusion, the POD-P stabilizer is more effective than the POD-Q to damp power oscillations. However, the grid-forming converters that use POD-P stabilizers require an energy buffer connected to the DC bus in order to exchange active power. While the STATCOM does not require any energy storage, but its damping effectiveness is much more limited. Anyway, a grid-forming converter with a POD-P stabilizer does not require a battery with a high energy capacity. According to the example of this section, power damping of a line transmitting 100 MW can be achieved through a net energy absorption of approximately 40 kWh that for a line of these characteristics is not an excessively high value.

8.6.3 Dynamic Response to a Line Tripping

This section shows the dynamic response of the two-area system when one of the two lines of the interconnection is tripped (see Figure 8.12). Before the tripping, each line transmitted 50 MW, but when the disturbance appears (at $t = 0$ s), the active power oscillates until reaching 100 MW in steady state, and it was the total power transmitted by the two lines before the tripping. Figure 8.18 shows the power oscillation through the healthy line after the tripping, in the cases of no compensation (top graph) and when using the POD-P and POD-Q stabilizers (bottom graph). In last two cases, power oscillation damping is achieved in less than 4 s.

During the line tripping event, the active and reactive power exchanges by the grid-forming and the STATCOM are shown in Figure 8.19. At the top, it is observed that the active power of the grid-forming converter varies between ±40 MW during the first oscillation. Likewise, at the bottom, corresponding to the STATCOM, the reactive power oscillates between ±20 MVar. In this case, it can be concluded that the STATCOM is more effective since it exchanges less power than the grid-forming converter while achieving nearly the same damping.

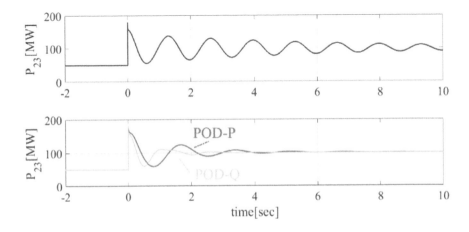

FIGURE 8.18   Active power transmitted by the interconnection during a line tripping.

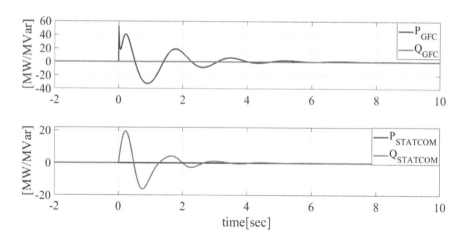

FIGURE 8.19 Active and reactive power exchange by the GFC and the STATCOM.

8.7 CONCLUSIONS

Nowadays, there is a growing concern about power system stability as a consequence of the increasing incorporation of renewable generation. Among stability issues, low-frequency oscillations arise as one of the most important concerns. Traditionally, these oscillations have been dampened by tuning the PSS of conventional SGs. However, modern PECs can contribute to provide this service efficiently. In this chapter, two stabilizers have been proposed; demonstrating the ability of both stabilizers to damp low-frequency oscillations in power systems. The first stabilizer, named POD-P, has been implemented in a grid-forming converter acting on the power angle, and consequently, exchanging active power to damp power oscillations. The second stabilizer, named POD-Q, has been implemented in a STATCOM using the internal voltage magnitude to damp system oscillations and exchanging only reactive power. Small-signal stability analysis has been carried out based on the linearized model of the system demonstrating the effectiveness of the proposed PODs. This effectiveness has also been proved through simulation, using comprehensive models of the nonlinear system. Simulation results, as well as the stability study, show the superiority of the POD-P stabilizer over the POD-Q, but at a cost of having to use some kind of energy supply in the DC bus to support the power interchange during system stabilization.

APPENDIX A. SYSTEM PARAMETERS AND OPERATION POINTS

TABLE 8.3
System Parameters

Parameter	Symbol	Value	Units
Stator resistance	R_s	0.0	pu
d-axis synchronous reactance	X_d	1.8	pu
q-axis synchronous reactance	X_q	1.7	pu
d-axis transient reactance	X'_d	0.3	pu
Line resistance	R_e	0.0	pu
Line reactance	X_e	0.3	pu
d-axis transient open-circuit constant	T'_{d0}	8.0	s
Inertia constant	H	6.175	s
Damping factor	D	0.0	pu
Exciter time constant	T_A	0.2	S
Exciter gain	K_A	200	pu
Frequency (50 Hz)	ω_s	314.16	rad/s
PEC reactance	X_g	0.15	pu
Washout time constant	$\mathbf{T}_w$	1.00	S
Washout gain	$\mathbf{K}_w$	3.00	pu
Droop gain	$\mathbf{R}_g$	0.05	pu

TABLE 8.4
State Variables and Algebraic Variables at the Operation Point when $V_\infty = 1$ pu, $V_t = 1$ pu, and $\theta = 9^0$

State Variables	Symbol	Value	Units
q-axis transient voltage	E'_q	0.8793	pu
q-axis angle	δ	48.647	degrees
Rotor speed	ω	314.159	rad/s
Excitation voltage	E_{fd}	1.4258	pu
Algebraic Variables	Symbol	Value	Units
d-axis current	I_d	0.3643	pu
q-axis current	I_q	0.3753	pu
d-axis voltage	V_d	0.6381	pu
q-axis voltage	V_q	0.7700	pu
Inputs	Symbol	Value	Units
Mechanical torque	T_m	0.5214	pu
Reference voltage	V_{ref}	1.0071	Pu

APPENDIX B. LINEARIZED MODELS

a. Single-machine infinite-bus model (baseline case):
Variable states are $\Delta x = \left[\Delta E'_q, \Delta \delta, \Delta \omega, \Delta E_{fd} \right]^t$, the vector of algebraic variables $\Delta y = \left[\Delta I_d, \Delta I_q, \Delta V_d, \Delta V_q \right]^t$. The input vector $\Delta u = \left[\Delta T_m, \Delta V_{ref} \right]^t$. The nonzero values of the matrices $A, B, C, D,$ and E are as follows:

MATRIX A [4 × 4]

$$A_{11} = -\frac{1}{T'_{d0}} \quad A_{14} = +\frac{1}{T'_{d0}} \quad A_{23} = \omega_s \quad A_{31} = -\frac{I_q^0}{2H} \quad A_{33} = -\frac{D}{2H} \quad A_{44} = -\frac{1}{T_A}$$

MATRIX B [4 × 4]

$$B_{11} = -\frac{X_d - X'_d}{T'_{d0}} \quad B_{31} = -\frac{I_q^0}{2H}\left(X_q - X'_d \right) \quad B_{32} = -\frac{1}{2H}\left[E_q^{'0} + \left(X_q - X'_d \right) I_d^0 \right]$$

$$B_{43} = \frac{K_A}{T_A}\left(\frac{V_d^0}{V_t^0} \right) \qquad B_{44} = \frac{K_A}{T_A}\left(\frac{V_q^0}{V_t^0} \right)$$

MATRIX C [4 × 4]

$$C_{21} = 1 \quad C_{32} = -V_\infty \cos\delta^0 \quad C_{42} = +V_\infty \sin\delta^0$$

MATRIX D [4 × 4]

$$D_{12} = X_q \qquad D_{13} = -1 \quad D_{21} = -X'_d \qquad D_{24} = -1$$

$$D_{32} = X_e \qquad D_{33} = +1 \quad D_{41} = -X_e \qquad D_{44} = +1$$

MATRIX E [4 × 2]

$$E_{31} = \frac{1}{2H} \quad E_{42} = \frac{K_A}{T_A}$$

b. SG and GFC connected to an infinite-bus:
In this model, the vector of variable states is extended with two additional states, $\Delta \delta_g$ and Δz. The nonzero elements of matrices A and B are the same as those indicated in the base case (single-machine infinite-bus) plus the following:

MATRIX A [6 × 6]

$$A_{52} = -R_g K_s \omega_s \quad A_{53} = -K_w \omega_s \quad A_{56} = +K_w \omega_s \quad A_{63} = \frac{1}{T_w} \quad A_{66} = -\frac{1}{T_w}$$

MATRIX B [6 × 4]

$$B_{53} = \left(\frac{E_g^0}{X_g K_s} \right)\left[\left(\frac{V_d^0}{V_t^0} \right) \sin\left(\delta^0 - \theta^0 \right) - 1 \right] \quad B_{54} = \left(\frac{E_g^0}{X_g K_s} \right)\left[\left(\frac{V_q^0}{V_t^0} \right) \sin\left(\delta^0 - \theta^0 \right) \right]$$

MATRIX C [4 × 6]

$$C_{21} = 1 \quad C_{32} = -\left(K_g E_g^0 \cos\left(\delta_g^0 - \delta^0\right) + K_e V_\infty \cos\delta^0\right) \quad C_{35} = K_g E_g^0 \cos\left(\delta_g^0 - \delta^0\right)$$

$$C_{41} = -K_d' \quad C_{42} = -\left(K_g E_g^0 \sin\left(\delta_g^0 - \delta^0\right) - K_e V_\infty \sin\delta^0\right) \quad C_{45} = K_g E_g^0 \sin\left(\delta_g^0 - \delta^0\right)$$

MATRIX D [4 × 4]

$$D_{12} = X_q \quad D_{13} = -1 \quad D_{21} = -X_d' \quad D_{24} = -1 \quad D_{32} = -K_d'\left(X_q - X_d'\right) \quad D_{33} = +1 \quad D_{44} = +1$$

MATRIX E [6 × 2]
The nonzero elements of matrix E are the same as in the baseline case.

c. SG and STATCOM connected to an infinite-bus:
In this model, the vector of variable states is extended with one additional state, Δz. The nonzero elements of matrix A are the same as those indicated in the baseline case (single-machine infinite-bus) plus the following:

MATRIX A [5 × 5]

$$A_{53} = \frac{1}{T_w} \qquad A_{55} = -\frac{1}{T_w}$$

MATRIX B [5 × 4]
The nonzero elements of matrix B are the same as in the baseline case.

MATRIX C [4 × 5]

$$C_{21} = 1 \quad C_{32} = -\left(K_g E_g^0 \cos\left(\theta^0 - \delta^0\right) + K_e V_\infty \cos\delta^0\right) \quad C_{33} = K_g K_w \sin\left(\theta^0 - \delta^0\right)$$

$$C_{35} = -K_g K_w \sin\left(\theta^0 - \delta^0\right) C_{41} = -K_d' \quad C_{42} = -\left(K_g E_g^0 \sin\left(\delta_g^0 - \delta^0\right) - K_e V_\infty \sin\delta^0\right)$$

$$C_{43} = -K_g E_g^0 \cos\left(\delta_g^0 - \delta^0\right) C_{45} = +K_g E_g^0 \cos\left(\delta_g^0 - \delta^0\right)$$

MATRIX D [4 × 4]
The nonzero elements of matrix D are the same as in the baseline case

MATRIX E [5 × 2]
The nonzero elements of matrix E are the same as in the baseline case

APPENDIX C. MODELS PARAMETERS

GENROU parameters:
$H = 6,3s, \quad D = 0, \quad Td0' = 6,47, \quad Td0'' = 0,022, \quad Tq0' = 0,61, \quad Tq0'' = 0,034$
$xd = 2,135, \quad xq = 2,046, \quad xd' = 0,34, \quad xq' = 0,573, \quad xd'' = xq'' = 0,269, \quad xl = 0,234$
$s1 = 0,1275, \quad s2 = 0,2706$

Excitation system model IEEE type ST1

$T_R = 0{,}01s$, $T_B = 10s$, $T_c = 1s$, $K_A = 200$, $T_A = 0s$ $Vi_{max} = 999$, $Vi_{min} = -999$, $V_{R,max} = 999$, $V_{R,min} = -999$, $K_c = K_F = 0$, $T_F = 1s$

Speed governor model and turbine IEEEG1

$K = 20$, $K_1 = 0{,}3$, $K_3 = 0{,}3$, $K_5 = 0{,}4$, $K_7 = 0$ $T_1 = T_2 = 0$, $T_3 = 0{,}1s$, $T_4 = 0{,}3s$, $T_5 = 7s$, $T_6 = 0{,}6s$, $T_7 = 0$

$K_2 = K_4 = K_6 = K_8 = 0$, $U_0 = 0.5$, $U_c = -0.5$, $P_{max} = 1$, $P_{min} = 0$

REFERENCES

[1] PES-TR77. (2020). "Stability definitions and characterization of dynamic behavior in systems with high penetration of power electronic interfaced technologies." Technical report. IEEE Power and Energy Society.
[2] Rocabert, J., Luna, A., Blaabjerg, F., & Rodriguez, P. (2012). "Control of power converters in AC microgrids." *IEEE Transactions on Power Electronics, 27*(11), 4734–4749.
[3] Yazdani, S., Ferdowsi, M., Davari, M., & Shamsi, P. (2019). "Advanced current-limiting and power-sharing control in a PV-based grid-forming inverter under unbalanced grid conditions." *IEEE Journal of Emerging and Selected Topics in Power Electronics, 8*(2), 1084–1096.
[4] D'Arco, S. & Suul, J. A. (2013). "Equivalence of virtual synchronous machines and frequency-droops for converter-based microgrids." *IEEE Transactions on Smart Grid, 5*(1), 394–395.
[5] Zhong, Q.-C., et al. (2013). "Self-synchronized synchronverters: Inverters without a dedicated synchronization unit." *IEEE Transactions on power electronics, 29*(2), 617–630.
[6] Rodriguez, P., et al. (2018). "Flexible grid connection and islanding of SPC-based PV power converters." *IEEE Transactions on Industry Applications, 54*(3), 2690–2702.
[7] Meng, X., Liu, J., & Liu, Z. (2018). "A generalized droop control for grid-supporting inverter based on comparison between traditional droop control and virtual synchronous generator control." *IEEE Transactions on Power Electronics, 34*(6), 5416–5438.
[8] Baltas, G. N., Lai, N. B., Marin, L., Tarraso, A., & Rodriguez, P. (2020). "Grid-forming power converters tuned through artificial intelligence to damp subsynchronous interactions in electrical grids." *IEEE Access, 8*, 93369–93379.
[9] Sauer, P. W., & Pai, M. A. (1998). *Power System Dynamics and Stability* (Vol. 101). Upper Saddle River, NJ: Prentice Hall.
[10] Xiong, L., et al. (2020). "Modeling and stability issues of voltage-source converter dominated power systems: A review." *CSEE Journal of Power and Energy Systems*, 1–18. https://ieeexplore.ieee.org/document/9265486

9 Grid-Forming Dynamic Stability under Large Fault Events – Application to 100% Inverter-Based Irish Power System

Xianxian Zhao and Damian Flynn
University College Dublin

CONTENTS

9.1 GRID-FORMING REQUIREMENTS FOR FUTURE GRIDS WITH HIGH SHARES OF RENEWABLES

For many relatively small and synchronously isolated power systems, such as Ireland and Great Britain, to achieve net-zero ambitions, the integration of high shares of inverter-based renewable energy sources (mainly wind and solar photovoltaic (PV)) is necessary. Maintaining such power systems at a (near) constant frequency and voltage, while ensuring high reliability levels is particularly challenging, since, amongst various reasons, the increased variability and uncertainty associated with wind and solar generation can lead to more frequent and larger voltage fluctuations, and challenges to generation-demand balancing.

The stability characteristics of synchronous generator-based power systems are well known and understood since they are constrained by the physical electromechanical properties of such machines. However, power systems with high inverter shares can introduce new issues of concern. Figure 9.1 presents some major challenges associated with high renewable systems, while indicating the ease with which they can be potentially addressed. The challenges are mainly due to (a) device models being hidden, and the resulting power system dynamic characteristics not being clearly defined, and (b) analysis methods to assess device and system behaviours still being in development [1,2]. Clearly, it follows that more detailed (and open) models are required, supported by enhanced device control methods, and more advanced stability assessment tools, including recognition of new phenomena, such as inverter control interactions [3].

At present, power electronic inverters, as typically seen in wind and PV farms, are mostly grid-following in nature, such that they follow the grid voltage angle using phase-locked loops (PLLs), and they behave as current sources. Hence, grid-following inverters cannot operate without synchronous machines (SGs) or synchronous

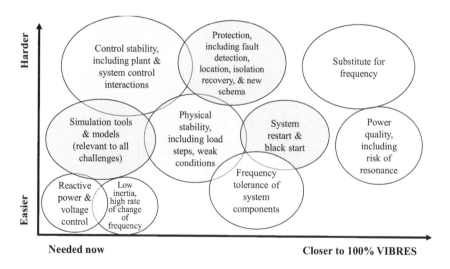

FIGURE 9.1 Stability challenges associated with achieving higher shares of inverter-based renewable energy sources [4].

condensers (SCs) nearby, since they need something to follow, and they require relatively stiff grids (rates of change of system states should be much slower than the PLL response speed) in order to achieve their control targets. However, with few or no SGs (or SCs) online nearby, the system states are likely to vary much more dynamically, and individual inverter PLLs may struggle to satisfactorily achieve the required tracking capability. This is particularly true for grid-following inverters connected at weak grid locations, leading to reduced transient performance and increased likelihood of system instability [5]. Therefore, grid-forming inverters (GFMIs), which are able to independently create the grid voltage frequency and angle without relying on nearby SGs or SCs, and which can provide an immediate response to system disturbances by behaving as voltage sources, are anticipated to offer a technical and economical solution towards maintaining the same reliability standards and stability levels as today [6]. It is noted that GFMIs are distinct from grid-supporting inverters, with the latter also being able to support the grid voltage and frequency, but they require nearby GFMIs, or SGs, since they don't independently generate their own voltage angle but instead employ PLLs to synchronise themselves with the grid. Consequently, they can also be categorised as being grid-following inverters.

As GFMIs are envisioned to replace synchronous machines, many different grid-forming controls have been proposed, mainly divided as droop control [7–9], mimicking synchronous machine operation by emulating a synchronous machine's physical dynamics [10], or reproducing the swing equation [11,12], synchronous machine matching control using the DC-link capacitor voltage to realise the swing equation [13–15], and virtual oscillator control [16,17]. Grid-forming pilot projects and field demonstrations are also emerging, most notably for battery energy storage systems [18] and wind turbines [19]. Given the likely increased importance of GFMIs for future systems, and particularly noting decarbonisation aspirations in many countries, it follows that power systems entirely based on power electronics inverters (wind, solar, battery, HVDC interconnection) may not be that far away. Consequently, GFMI dynamic stability under large fault events is of interest, and a reduced Irish transmission grid with 100% power electronic inverters, without any synchronous machines, is considered as a test case. In comparison to synchronous generators, GFMIs exhibit much superior voltage and frequency regulation capability, if the energy input and inverter overcurrent capability are assumed ideal, due to the fast and full controllability of voltage source inverters [8,10]. However, for fault events, the limited overcurrent capability (2~3 times smaller than SGs) of GFMIs is an issue, and instability problems can result if the current limiting control is not suitably designed [20]. The dynamic stability of GFMIs under "hard" current reference (with [21], or without [22,23], anti-windup integration for the outer voltage loop, using active current prioritisation [22], or scaling current approach [21,23]), "soft" virtual impedance (VI) current limiting [24], or an intricate projected method (where the real-time grid short-circuit ratio needs to be estimated [25]) has been extensively studied for balanced faults. A "soft" VI current limiter is required for GFMIs, since it not only makes the inverter more robust for both fault and downward phase jump disturbances, as compared to "hard" direct current reference limiting but also avoids high-frequency oscillations, which can be easily excited by active or reactive current prioritisation current reference limiting. Under both fault and grid phase jump events, while ensuring strict current limiting, a combination of both hard and soft current limiting measures has also been proposed for GFMIs to aid their transient stability [26]. Moreover, virtual angular speed freezing has been proposed for GFMIs in order to enhance transient stability under fault conditions. Given that wind and solar farms will most likely be in different locations to existing synchronous machines, the rating and location of GFMIs have been shown to be critical to securely operating the grid in the presence of both grid-forming and grid-following inverters [27,28], while distributing GFMIs at individual buses across the system can reduce GFMI requirements [29,30]. However, although GFMI transient stability under fault conditions has been extensively studied, case studies based on real systems and associated validations remain few.

The remainder of the chapter is organised as follows: droop, virtual synchronous machine (VSM) and dispatchable virtual oscillator (dVOC) control for GFMIs are introduced, supported by current reference and VI current limiting control, and virtual angular speed freezing, in Section 9.2; *Urban* and *Remote* scenarios for inverter locations for a future Irish transmission system are considered in Section 9.3, and an array of case studies are presented, including the *Urban* Irish grid with 100% GFMIs (all droop (or VSM) control, all dVOC control, or a mix of both), with hybrid current limiting and without virtual angular speed freezing control, and case studies of the *Remote* Irish grid with 100% GFMIs under either current reference or VI current limiting control, and with/without virtual angular speed freezing; finally, Section 9.4 draws conclusions for operating and controlling 100% power electronic-based power systems.

9.2 GFMI MODELLING, CURRENT LIMITING, AND CONTROL DESIGN

Droop, VSM, and dVOC control-based GFMIs, with inner cascaded voltage and current proportional integral (PI) control, together with a number of current limiting control schemes (active and reactive priority, scaled current reference and VI current limiting) are introduced, while a virtual angular speed freezing technique is presented to enhance transient stability while being insensitive to phase jumps. Synchronous machine matching control using the DC-link capacitor voltage to realise the swing equation [13–15] has been shown to cause large deviations in the DC-link voltage (which needs to be tightly controlled, notably for HVDC and wind turbine applications), and hence is not studied here.

9.2.1 DROOP CONTROL WITH INNER CASCADED VOLTAGE AND CURRENT PI CONTROL LOOPS

A voltage source inverter with a PI-controlled DC current source, capacitor, and inductance-capacitor-inductive (LCL) filter is shown in Figure 9.2 connected to an AC grid. The DC current source represents the DC-side input power, perhaps from a PV panel, wind turbine, or energy storage system. The AC grid is represented by an ideal voltage source in series with an impedance (R_g, L_g), while a step-up transformer, R_c, L_c, represents the grid-side *RL* filter. Figure 9.2 also indicates a droop-based control

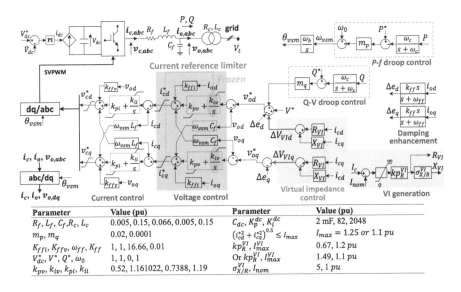

Parameter	Value (pu)	Parameter	Value (pu)
R_f, L_f, C_f, R_c, L_c	0.005, 0.15, 0.066, 0.005, 0.15	$C_{dc}, K_p^{dc}, K_i^{dc}$	2 mF, 82, 2048
m_p, m_q	0.02, 0.0001	$\left(i_{cd}^{*2} + i_{cq}^{*2}\right)^{0.5} \leq I_{max}$	$I_{max} = 1.25\ or\ 1.1\ pu$
$K_{ffi}, K_{ffv}, \omega_{ff}, K_{ff}$	1, 1, 16.66, 0.01	kp_R^{VI}, I_{max}^{VI}	0.67, 1.2 pu
$V_{dc}^*, V^*, Q^*, \omega_0$	1, 1, 0, 1	Or kp_R^{VI}, I_{max}^{VI}	1.49, 1.1 pu
$k_{pv}, k_{iv}, k_{pI}, k_{iI}$	0.52, 1.161022, 0.7388, 1.19	$\sigma_{X/R}^{VI}, I_{nom}$	5, 1 pu

FIGURE 9.2 Grid-forming inverter connected to a DC current source and an AC grid via a LCL filter under droop-based control, with a combination of virtual impedance and saturation current limiting, and a damping enhancement block, and its control structure and parameters.

structure, with cascaded voltage and current PI decoupling control loops. The active power/frequency droop control generates the inverter internal voltage angular speed, ω_{vsm}, and angle, θ_{vsm}. The reactive power/voltage droop control generates the voltage deviation, ΔV, which is then combined with V^* to form the d-axis output voltage reference, v_{od}^*. The q-axis output voltage reference, v_{oq}^*, is set to zero, such that the dq frame for the decoupling control rotates at speed ω_{vsm}, with the d-axis aligned to the internal/output voltage across the LCL filter capacitor. A cascaded voltage and current PI control structure has been adopted here since the LC filter dynamics can be fully exploited to achieve the GFMI control requirements and current limiting. The PI controller parameters are based on [31], which are tuned to provide the largest damping ratio for a GFMI connected to a grid with a high short-circuit ratio. Damping enhancement control (Δe_d and Δe_q are the high-pass filter output of the output current i_{od} and i_{oq}) is also incorporated to the output voltage references, v_{od}^* and v_{oq}^*.

Threshold VI current limiting is also shown in Figure 9.2, whereby if the measured inverter current magnitude, I_c, exceeds the specified value, I_{nom}, the virtual resistance and reactance, $R_{VI} > 0$ and $X_{VI} > 0$, and the VI control are activated. The generated dq-axis voltage drops, ΔV_{VId} and ΔV_{VIq}, are then combined with v_{od}^* and v_{oq}^*. The VI control parameters (kp_R^{VI} and $\sigma_{X/R}^{VI}$) are tuned to limit the inverter current to I_{max}^{VI} when a bolted three-phase fault is applied at the capacitor filter terminals, based on the tuning principles described in [24].

The control logic for the scaled current reference limiting control is shown in Figure 9.3, such that when $I_c^{*0} \geq I_{max}^{sat}$, i_{cd}^{*0} and i_{cq}^{*0} are scaled to ensure that $i_{cd}^{*2} + i_{cq}^{*2} \leq I_{max}^2$,

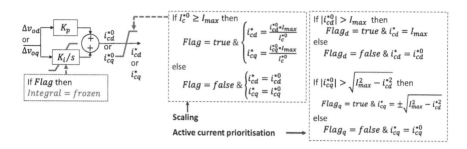

FIGURE 9.3 Current scaling and active current prioritisation saturation limiter including anti-windup integration in outer voltage PI control, where i_{cd}^{*0} and i_{cq}^{*0} are the d-axis and q-axis current references from the outer voltage control loop, and I_c^{*0} is the relevant amplitude.

while, the integrator of the outer PI controller is frozen by setting its input to zero. Figure 9.3 also shows the active current (d-axis) prioritised current reference limiting, while reactive current (q-axis) prioritised limiting is the same as its active current counterpart with d and q subscripts exchanged (further details in [32,33]). It is noted that alternative strategies can achieve the same mathematical relationship, $\sqrt{i_{cd}^{*2} + i_{cq}^{*2}} \le I_{max}^2$, for example, $i_{cd}^* = I_{max} \cos(\theta)$, $i_{cq}^* = I_{max} \sin(\theta)$, where θ is a fixed angle, as in [34].

If only current reference limiting control is applied, it is likely that the GFMI in Figure 9.2 will quickly lose stability or output high-frequency power oscillations following a large fault disturbance or downward phase jump (GFMI stability under upward phase jumps is of little concern [26]). Inverter stability can be improved if "soft" VI current limiting control is incorporated, since when VI is activated, the outer voltage PI control operates as normal, and the output voltage can be directly controlled. However, the initial inverter fault current can still be high, with a decay time of ≈ 20 ms. Consequently, a combination of VI and scaled current reference saturation control is proposed. In comparison with the current reference saturation strategy implemented in [34], scaling acts in proportion to the current magnitude, such that it is not necessary to define a fixed current reference angle. The current limit for current saturation control is set higher than that for VI control, $I_{max}^{VI} < I_{max}$, for example, 1.25 pu relative to 1.2 pu. Thus, scaled current reference control tends to be active during the initial transients, while VI control dominates the "longer term" GFMI dynamics.

9.2.2 VIRTUAL SYNCHRONOUS MACHINE CONTROL

VSMs, also known as virtual synchronous generators (VSGs), or synchronverters, have been widely studied. A synchronverter [10] emulates the electrical behaviour of a synchronous machine, while a VSM [11,12] reproduces the swing equation to generate the voltage reference, as shown in Figure 9.4a. However, a fixed $\omega_0 = 1$ pu is applied here, instead of an estimated frequency from a PLL, as in [11], noting that a GFMI should independently form the grid voltage angle and avoid fundamental

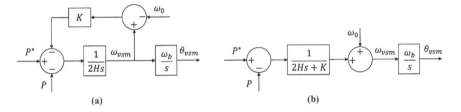

FIGURE 9.4   Virtual synchronous machine control, (a) is equivalent to (b).

shortcomings of grid-following inverters. Figure 9.4a matches Figure 9.4b, since the former can be formulated and transformed as

$$\omega_{vsm} = \frac{1}{2Hs}\left(P^* - P - K\left(\omega_{vsm} - \omega_0\right)\right), \tag{9.1}$$

$$\left(2Hs + K\right)\omega_{vsm} = P^* - P - K * \omega_0, \tag{9.2}$$

$$\omega_{vsm} = \frac{P^* - P}{2Hs + K} - \frac{K}{2Hs + K}\omega_0 = \frac{P^* - P}{2Hs + K} - \omega_0, \tag{9.3}$$

Equation (9.3) is shown in Figure 9.4b, such that Figure 9.4a and b are equivalent.

The VSM control in Figure 9.4a and b is the same as the P/f droop control in Figure 9.2, which can be expressed and transformed as follows

$$\omega_{vsm} = m_p\left(P^* - \frac{\omega_c}{s + \omega_c}P\right) + \omega_0, \tag{9.4}$$

$$\omega_{vsm} = m_p\left(\frac{\omega_c}{s + \omega_c}\left(P^* - P\right)\right) + \omega_0 = \frac{1}{\frac{1}{m_p\omega_c}s + \frac{1}{m_p}}\left(P^* - P\right) + \omega_0. \tag{9.5}$$

Therefore, the form of equations (9.3) and (9.5) are equivalent, noting that

$$H = \frac{1}{2m_p\omega_c}, K = \frac{1}{m_p}. \tag{9.6}$$

As will be shown later in Section 9.3, the VSM and droop controls create the same simulation results, subject to the parameter selections defined in equation (9.6), and other control aspects being unchanged. When designing a robust droop control implementation, a large number of simulation studies suggest that the droop coefficient should be selected as $m_p \leq 0.02$ [31]. A low value for m_p implies a strong load-sharing contribution for a GFMI, while recognising that $m_p = 0.04$ or 0.05 is typical for thermal (synchronous machine) power plants. With m_p (or damping coefficient K) chosen, ω_c

(or H) must also be selected. Small H improves small-signal stability, or small-disturbance oscillatory stability, with power oscillations quickly damped. In contrast, for large H, more "inertial" energy is stored, and the (virtual) rotor speed will deviate more slowly under large disturbances, such as severe faults and generator trips.

9.2.3 DISPATCHABLE VIRTUAL OSCILLATOR CONTROL-BASED GRID-FORMING INVERTER MODELLING

Dispatchable virtual oscillator (dVOC) control adopts a different approach to generating the voltage amplitude and angle references, by recognising the couplings between active and reactive power [17]. Both droop (or VSM) and dVOC controls are investigated under fault conditions in Section 9.3. The implemented dVOC control strategy follows the same principles as those shown in Figure 9.2, except that the virtual angle, θ_{vsm}, and dq-axis voltage references, v_{od}^* and v_{oq}^*, are generated as

$$\frac{1}{\omega_b}\begin{bmatrix} \dot{u}_{o\alpha}^* \\ \dot{u}_{o\beta}^* \end{bmatrix} = \eta\left(K\begin{bmatrix} u_{o\alpha}^* \\ u_{o\beta}^* \end{bmatrix} - R(\rho)\begin{bmatrix} i_{o\alpha} \\ i_{o\beta} \end{bmatrix} + \alpha\left(1 - \frac{U_o^{*2}}{V_{ref}^2}\right)\begin{bmatrix} u_{o\alpha}^* \\ u_{o\beta}^* \end{bmatrix}\right),$$

$$K = \frac{1}{V_{ref}^2}R(\rho)\begin{bmatrix} P_{ref} & Q_{ref} \\ -Q_{ref} & P_{ref} \end{bmatrix}, \tag{9.7}$$

$$\theta_{vsm} = \omega_0 t + \tan^{-1}\left(\frac{u_{o\beta}^*}{u_{o\alpha}^*}\right), \omega_{vsm} = \dot{\theta}_{vsm}/\omega_b, \tag{9.8}$$

$$v_{od}^* = \sqrt{u_{o\alpha}^{*2} + u_{o\beta}^{*2}} - \Delta V_{VId} - \Delta e_d, v_{oq}^* = 0 - \Delta V_{VIq} - \Delta e_q, \tag{9.9}$$

where, $R(\rho_i) = \begin{bmatrix} \cos(\rho_i) & -\sin(\rho_i) \\ \sin(\rho_i) & \cos(\rho_i) \end{bmatrix}$, $\rho_i = \frac{l_i}{r_i}$, and l_i and r_i are the transmission line inductance and resistance. Here, ρ_i is assumed to be the same for all transmission lines (in [35], simulation and experimental results both demonstrate that the system remains stable even when this assumption is not true). In order to achieve a fair comparison, the "soft" VI current limit and damping enhancement controls shown in Figure 9.2 are also applied to v_{od}^* and v_{oq}^* in equation (9.3). Details of the dVOC controls are shown in Figure 9.5, with $\eta/V_{ref}^2 = m_p$, $\alpha V_{ref} = 1/m_q$, such that the dVOC steady-state active and reactive power droop settings are equivalent to the previous droop control settings (whose steady-state droop characteristic can be found in [36]). In addition, for the case studies shown in Section 9.3, the voltage set-points and active and reactive power set-points for both controls are unchanged.

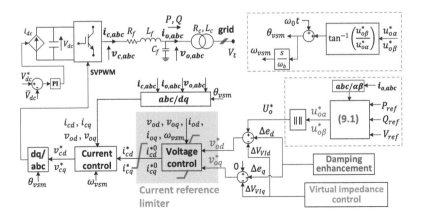

FIGURE 9.5   Grid-forming inverter under dispatchable virtual oscillator control, combined with virtual impedance and current saturation limiters, and a damping enhancement block.

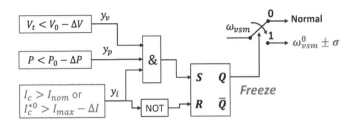

FIGURE 9.6   Unified virtual angular speed freezing for a grid-forming inverter system without synchronous generators (condensers), where V_0, P_0, and ω_{vsm}^0 are the pre-fault terminal voltage, active power output, and virtual angular speed, and ΔV, ΔP, and ΔI are small constants, set here as 0.25, 0.2, and 0.05 pu. σ is a small positive constant, giving the GFMI additional freedom (e.g. small negative σ can speed up GFMI post-fault recovery for $P_0 > 0$ and very weak connections), and set as 0 here.

9.2.4   GRID-FORMING INVERTER ANGULAR SPEED FREEZING

The transient stability of GFMIs is affected by the transient evolution of the voltage angle, and the d-axis and q-axis components of the generated voltage. Since a power system may consist of many GFMIs, each employing different control strategies, and utilising different parameters, analysis of such systems can be challenging. Here, instead of controlling the virtual angular speed, or the associated angle, for each GFMI, it is proposed that their virtual angular speeds are frozen to their pre-fault values when a fault is detected and the inverter current exceeds a certain value (see Figure 9.6). It follows that the relative angular differences between the GFMIs after a fault is cleared should be relatively similar to the pre-fault values (variations will occur due to specific system conditions and parameter settings). Considering the power-angle relationship and equal area criterion, if the relative angle between voltage sources doesn't change too much during a fault, then the system is more likely to remain stable, and since all the GFMIs are frozen to the same angular speed, it follows that the fault critical clearing time should be increased. The freezing action is activated when both the output voltage and active power (V and P) are lower than the defined thresholds, V_l and P_l. Including the active power, P, input avoids the freezing action being triggered during grid phase jump down events. Finally, it is noted that no communication is required to implement the freezing technique, as seen in Figure 9.6, but the frozen angular speed input can be updated, as required, as system conditions change, using (slow) communications, or otherwise.

The virtual angular speed freezing technique of Figure 9.6 is formulated as (9.10)

$$\omega_{vsm} = \begin{cases} \text{normal,} & \text{Otherwise} \\ \omega_{vsm}^0, & \text{fault } \& \left(I_c > I_{nom}\right) \text{ or fault } \& \left(I_c^{*0} > I_{max} - \Delta I\right) \end{cases} \quad (9.10)$$

where, if only current reference saturation limiting is used, the logic $I_c^{*0} > I_{max} - \Delta I$ is chosen, otherwise $I_c > I_{nom}$ (i.e. VI is activated).

9.3   IRISH GRID CASE STUDY WITH 100% GFMIs

The transient dynamics of a reduced Irish transmission system based on the 2015 Irish winter peak demand with all generation being based on GFMIs is considered in order to assess some of the stability issues associated with future 100% renewable (wind and solar) power systems. For the test system, individual GFMIs can either employ droop (or VSM) control or dVOC control, with options for VI and scaled current reference saturation control, and virtual angular speed freezing is also considered. Electromagnetic transient (EMT) simulations are performed under the Dymola environment using the Modelica modelling language [37], whereby all modelling details are completely transparent and user-created models can be easily integrated. The focus here is placed on the transient stability impact of replacing synchronous generators by GFMIs on a test system, but relevant theoretical analysis on single-machine infinite bus system has been considered elsewhere [26,38].

9.3.1   URBAN AND REMOTE GRID MODELS

The Ireland and Northern Ireland transmission system is synchronously isolated from Great Britain and the rest of the European continental system, with operating voltage levels of 400, 275, 220, and 110 kV [39]. Most of the conventional generating fleet is connected at 220 kV and above voltage levels, although the majority of the network consists of 110 kV circuits spread across the entire system. Ireland is targeting up to 80% renewable generation by 2030, which implies that for ≈50% of the time, wind and solar availability will exceed the instantaneous demand [40]. However, due to system stability and security concerns, a system non-synchronous penetration (SNSP) upper limit of 75% is currently imposed, which is defined as follows:

$$SNSP = \frac{P_W + P_{HVDC(imp)}}{P_L + P_{HVDC(exp)}} \quad (9.11)$$

where, P_W, P_L, $P_{HVDC(imp)}$, and $P_{HVDC(exp)}$ are the actual wind (and solar) power, load demand, and power imported/exported through HVDC links, respectively. The system operators, EirGrid and SONI, have proposed pushing the SNSP ceiling towards 95% and beyond [40], but for a variety of technical reasons, operating the system at 100% SNSP is not yet considered viable, even if the instantaneous wind (and solar PV) share at any point in time could enable this to be achieved. Consequently, the Irish grid, which is synchronously isolated, and already accepting high SNSP levels, can be considered as a good starting point for investigating plausible 100% inverter-based systems without excessive network and/or load/generation growth changes being required. However, the operational rules for a 100% inverter-based system are likely to be noticeably different.

For systems with high inverter shares, particular focus must be placed on network faults, given the limited overcurrent capability of inverters, while also noting that they tend to be 2~3 orders of magnitude smaller in power rating relative to synchronous generators. Since three-phase faults generally represent a worst-case scenario, they are considered here for a 100% inverter-based Irish grid.

Figure 9.7a shows the reduced Urban Irish transmission grid, with important aspects of the simulation setup summarised:

- Only the 400, 275, and 220 kV transmission system voltage levels are represented, with lower voltage levels represented by equivalent loads. In total, 82 bus nodes, 47 constant impedance loads (actual and equivalent), 14 inverters, 110 transmission lines (equivalent π model), and multiple shunt capacitors and transformers are included.
- The 2015 Irish winter peak demand is applied, with ≈6 GW and ≈1.3 GVAr of active and reactive loads, acknowledging relatively flat demand growth in Ireland in recent years.

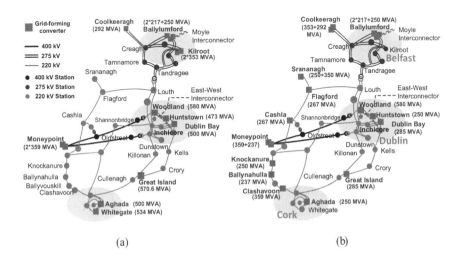

(a)   (b)

FIGURE 9.7   Single line diagram of (a) *Urban* and (b) *Remote* Irish grids with 100% grid-forming inverters, with bus nodes roughly corresponding to geographical locations.

- For the *Urban* configuration, the inverters are placed at the same locations as the critical set of online conventional units, based on existing operational (steady-state network loading and stability) rules for the Irish power system [41]. Such an approach implies that system-level issues, such as inter area flows, line overloading, voltage support, and system reserves, which may result due to changes in size and location of (renewable-based) inverters displacing fossil-fired synchronous generators are largely avoided. Instead, the focus is placed on the change (from synchronous machine to inverter) in the generation sources themselves.
- Individual inverter(s) are assumed to be large in size, with the same rating as the existing conventional units in order to reduce the number of parameters being modified between the existing and 100% inverter-based systems.

Figure 9.7b presents the *Remote* grid configuration, whereby the inverters are placed in more remote parts of the network (and at lower voltage levels), as might be expected in a variable renewables-dominated system [42]. Additional 110 kV lines (not shown in Figure 9.7b) are included in the western parts of the system to connect the new inverters to the transmission network. Both grids possess the same inverter capacity and load demand, but, for the *Remote* system, additional capacitors are added, as necessary, at inverter buses in the east and west, to maintain system voltage levels within acceptable levels.

9.3.2 *Urban* Irish Grid with 100% GFMIs (Droop, dVOC Control)

Three-phase faults are applied at all buses of the reduced *Urban* Irish grid in Figure 9.7a, assuming that all the inverters are grid forming, and that they are operating under either droop control or virtual oscillator control. Three buses are selected to present the results: Woodland, Kilroot and Great Island, located near Dublin, Belfast, and the rural region between Dublin and Cork, respectively. These locations are chosen for two main reasons: proximity to the major load centres and proximity to the inverter locations.

Initially it is assumed that all the GFMIs are based on droop control, with the combination of VI and scaling current reference saturation control, and imposed current limits of 1.2 and 1.25 pu, respectively. A three-phase, 200 ms bolted fault is applied at the Woodland (near Dublin) bus. Figure 9.8a shows the voltages at the three selected buses, with the voltage at Woodland falling close to zero, since it is close to the fault location, while the voltage dips at the other two, more remote, buses are not as large, with Kilroot being further away than Great Island. After clearing the fault, the voltage recovers quickly at all the buses, and the oscillations dissipate after less than 20 ms. Figure 9.8b and c shows the reactive (and active) power output immediately increasing (and reducing) after the fault occurs, with steady-state levels achieved within approximately 250 ms. The largest reduction in voltage is seen at the Woodland bus, given that it is closest to the fault location. Figure 9.8d shows the injected current from the inverters, which is limited in steady state below the 1.2 pu limit (due to parameter tuning of the VI current limiting control), even for the Woodland inverter. However, the injected current initially exceeds 1.2 pu (but still

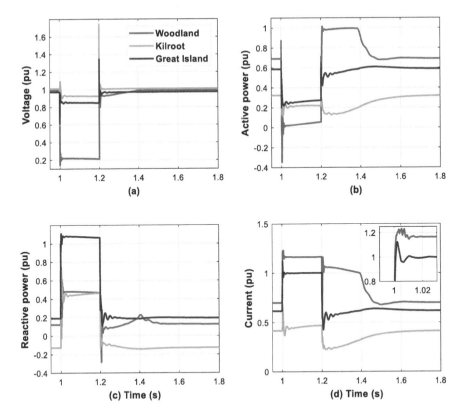

FIGURE 9.8 Voltage, active and reactive power output, and injected current of three GF inverters near Dublin, Belfast, and Cork ((a), (b), (c) and (d), respectively) for a 200 ms, three-phase bolted fault at the Woodland bus of the *Urban* Irish grid with 100% grid-forming inverters under droop control (combination of virtual impedance and current scaling limiting applied).

less than 1.25 pu) when the fault occurs, with the current saturation control strictly limiting the current within 1.25 pu.

In order to confirm that the "hard" current saturation control, when combined with VI current limiting control, does not negatively impact system stability, Figure 9.9 shows the case when only "soft" VI current limiting is implemented. The results in Figure 9.9a–d are very similar to those in Figure 9.8a–d, except that the current for the Woodland inverter transiently exceeds 1.4 pu and oscillates for ~20 ms (confirming that the VI control response time is too slow to limit the initial current, which involves measuring the current before modifying the filter capacitor voltage references. The outer voltage control loop slows the response). Given that similar system behaviour is achieved with/without current saturation control, the combined hybrid current limiting option is adopted for subsequent simulations of the *Urban* Irish grid.

The inverter control strategy is now switched from droop control to dVOC control, with the same fault conditions and fault location as above considered in Figure 9.10. The results are similar to Figure 9.8, mainly because: (a) for an inductive network,

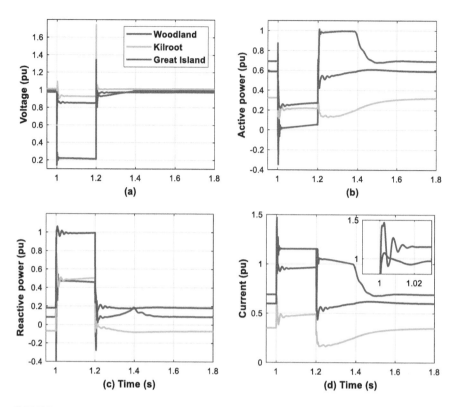

FIGURE 9.9 Voltage, active and reactive power output, and injected current of three GF inverters near Dublin, Belfast, and Cork ((a), (b), (c) and (d), respectively) for a 200 ms, three-phase bolted fault at the Woodland bus of the *Urban* Irish grid with 100% grid-forming inverters under droop control (virtual impedance current limit only applied).

dVOC control resembles droop control [17], (b) the steady-state expressions for the outer loop voltage references, and virtual angular speed and angle, are tuned to provide similar performance at nominal voltage, and (c) the same current limiting strategies are applied, which, when active, dominate the transient response. It follows that the nature of the current limiting strategies should form an important aspect of assessing the transient stability of GFMI-based systems. In comparison with droop control, the dVOC approach achieves a faster recovery period, given that the frequency and voltage responses are directly coupled, which leads to a reduction in the virtual frequency deviation (as seen in Figure 9.11) and, a smaller increase in the virtual angle during the fault.

To confirm that the VSM control in Figure 9.4 is the same as the droop control in Figure 9.2, VSM control is now applied for the same fault condition, with $H = 0.7962$ s and $K = 50$ in order to satisfy Equation 9.6. The simulation results are shown in Figure 9.12, with Figure 9.8 (droop control) superimposed. The similarity of the active power and injected current traces demonstrates that VSM control is equivalent to droop control.

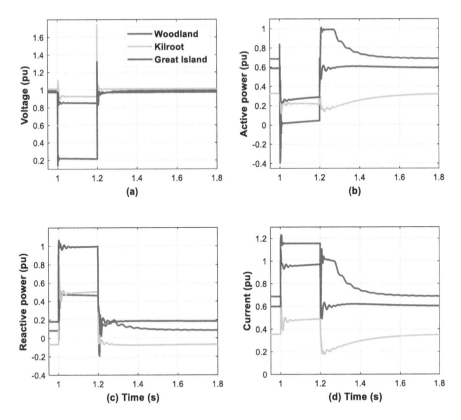

FIGURE 9.10 Voltage, active and reactive power output, and injected current of three GF inverters near Dublin, Belfast, and Cork ((a), (b), (c) and (d), respectively) for a 200 ms, three-phase bolted fault at the Woodland bus of the *Urban* Irish grid with 100% grid-forming inverters under dVOC control (combination of VI and saturation current limiting controls applied).

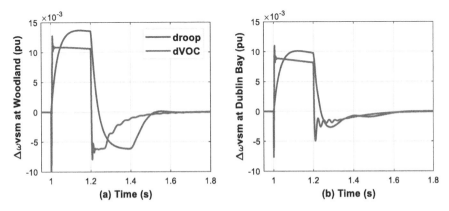

FIGURE 9.11 Virtual angular speed deviation, $\Delta\omega_{vsm}$, for two GF inverters at Woodland and Dublin Bay near Dublin ((a) and (b)) for a 200 ms, three-phase bolted fault at the Woodland bus of the *Urban* Irish grid with 100% grid-forming inverters under droop or dVOC control.

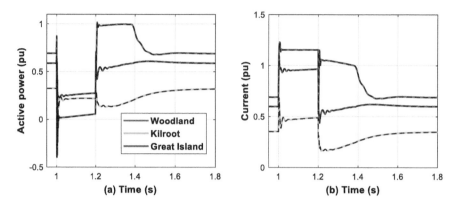

FIGURE 9.12 Active power output and injected current for three grid-forming inverters near Dublin, Belfast, and Cork ((a) and (b)) for a 200 ms, three-phase bolted fault at the Woodland bus of the *Urban* Irish grid with 100% grid-forming inverters under VSM control (droop control shown using dashed lines).

Finally, instead of all the GFMIs adopting either a droop or dVOC control strategy, dVOC control is arbitrarily applied to the inverters at the Aghada, Great Island, Huntstown, Woodland, and Coolkeeragh buses, and also one of the inverters at Moneypoint and Ballylumford, while the remaining inverters employ droop control to create a "mixed" control strategy. For the same fault conditions and fault location as before, Figure 9.13 shows similar performance to Figures 9.8 and 9.10, indicating that the grid-forming control scheme chosen has minimal impact on the overall response. However, this conclusion assumes that each control strategy applies the same current limiting strategy, maximum current limit, and DC-side energy sources.

For the droop, dVOC and mixed control strategies, the system quickly recovers its initial state once the fault clears, with no/low oscillations observed during and postfault. Similar results are achieved when the same fault is applied at all buses, which can be seen in [29]. It can be concluded that the system is robust for a GFMI-only system.

9.3.3 *REMOTE* IRISH GRID WITH 100% GFMIs (CURRENT LIMITING STRATEGIES AND VIRTUAL ANGULAR SPEED FREEZING)

Following on from the previous section, VI and (active and reactive current prioritised, scaled) current saturation current limiting with high/low limiting, with/without virtual angular speed freezing are now examined for the *Remote* grid, Figure 9.7b.

9.3.3.1 Current Saturation and Virtual Impedance Limiting Control

Case 1: 200 and 260 ms bolted three-phase faults are applied at the Woodland bus in the *Remote* system, with droop-based GFMIs under VI control and a current limit of 1.2 pu, *VI – 200 ms* and *VI – 260 ms*. Figure 9.14d shows that the "steady-state" injected current during the fault for the GFMI at Woodland remains within 1.2 pu, although the initial transient exceeds 1.4 pu. Compared to the GFMI at Woodland in

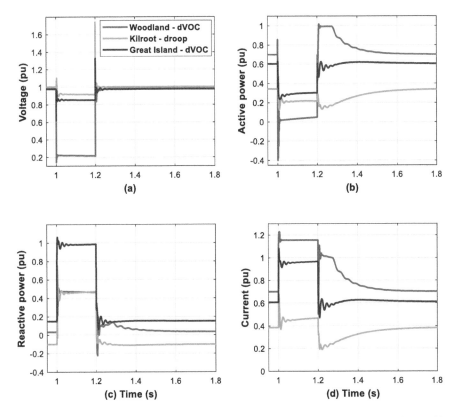

FIGURE 9.13 Voltage, active and reactive power output, and injected current for three grid-forming inverters near Dublin, Belfast, and Cork ((a), (b), (c) and (d), respectively) for a 200 ms, three-phase bolted fault at the Woodland bus of the *Urban* Irish grid with 100% grid-forming inverters for a mix of droop and dVOC controls (combination of VI and saturation current limiting controls applied).

Figure 9.8, Figure 9.14a–c shows that the GFMI under the *VI – 200 ms* scenario remains stable but takes longer to recover to its pre-fault state, indicating that the *Remote* system is less robust than the *Urban* configuration, given that much of the generation is now located in the west while the major loads are still located in the east. Comparing the two fault durations for the *Remote* system, Figure 9.14 shows that the GFMI takes longer to recover for the longer fault, given that the virtual angle is continuously increasing during the fault.

Case 2: With the droop-based GFMIs operating under current saturation control based on active, reactive current prioritisation or scaling, and the same current limit of 1.2 pu, a 260 ms, three-phase bolted fault is applied at the Woodland bus, as in Case 1. Figure 9.15 shows the response of the inverter at Woodland, with the inverter current being strictly limited within 1.2 pu under all three current reference saturation limiting approaches (initial current slightly exceeds 1.2 pu for ≈ 6 ms due to the one-step time delay (0.1 ms), implemented to represent signal measurement), Figure 9.15d. Under reactive current priority, large oscillations are observed

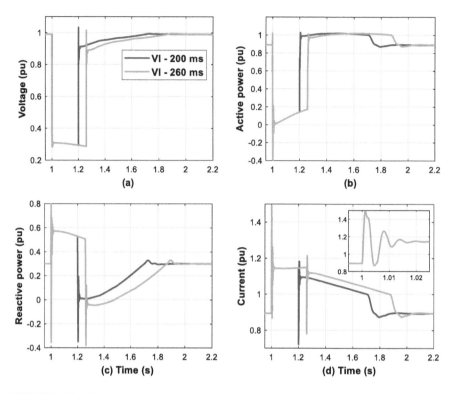

FIGURE 9.14 Voltage, active and reactive power output, and injected current for the grid-forming inverter at the Woodland bus ((a), (b), (c) and (d), respectively) for 200 and 260 ms, three-phase bolted faults at the Woodland bus of the *Remote* Irish grid (Case 1), with 100% droop control grid-forming inverters incorporating virtual impedance control with $I_{max}^{VI} = 1.2$ pu.

during the fault in the voltage, and active and reactive power output, for the GFMI. Under active current priority, the Woodland GFMI is unstable and doesn't return to its pre-fault state, since the injected current remains saturated after fault clearance, Figure 9.15b. Active current prioritisation also leads to large voltage and power oscillations during and post-fault clearance (see [43]), if the one-step time delay is removed (the time delay helps to smooth the inverter dynamics, and oscillations are largely avoided in Figure 9.15). In both scenarios, the high-frequency oscillations are excited by the LC filter, which is induced by the generated distorted d- and q-axis current references under active/reactive current prioritisation. Incorporating a resistor in series with the filter capacitor, C_f, helps to reduce the oscillations, but the frequency of the oscillations depends on the loading and system conditions, which makes it challenging to robustly design an appropriate filter. Furthermore, since active and reactive current prioritisation are not well aligned with the need to regulate both the active and reactive power, it follows that current saturation control based on active and reactive current prioritisation is not recommended for GFMIs. Under the scaling strategy, Figure 9.15a–c shows that the inverter is unstable, but the high-frequency

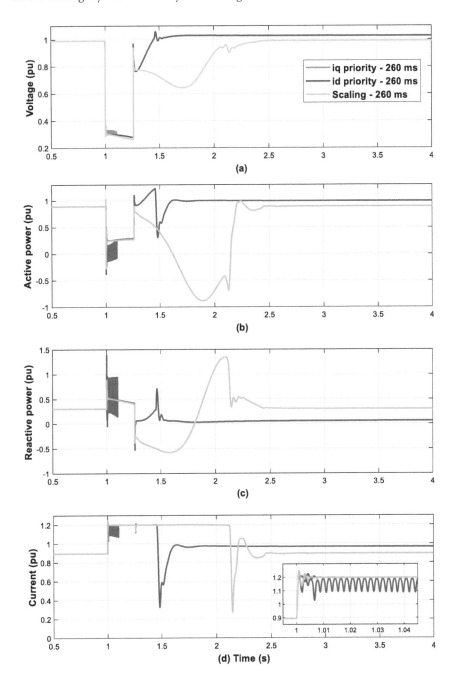

FIGURE 9.15 Voltage, active and reactive power output, and injected current for the grid-forming inverter at the Woodland bus ((a), (b), (c) and (d), respectively) for a 260 ms, three-phase bolted fault at the Woodland bus of the *Remote* Irish grid (Case 2), with grid-forming inverters applying active, reactive current priority, or scaling current saturation limiting with $I_{max} = 1.2$ pu.

oscillations are not that severe, and the current returns to its pre-fault state $\approx 1.2\,\text{s}$ after fault clearance, Figure 9.15d, which confirms that current scaling can be combined with VI current limiting to strictly bound the GFMI current. In comparison to *VI – 260 ms* in Case 1, which is stable, the unstable scenarios in Case 2 indicate that VI control is more effective than current saturation limiting control in ensuring GFMI transient stability for faults.

Case 3: Three-phase faults are now considered at a different location, of a longer duration, and adopting an alternative current limiting approach for the GFMIs, with $I_{max}^{VI} = I_{max} = 1.1$ pu, as follows: (a) 260 ms, three-phase bolted fault at Aghada (near Cork) under VI control (*VI – 260 ms*); (b) same as (a) but for a 305 ms fault (*VI – 305 ms*); and (c) scaled current limits for a 200 ms fault (*Scaling – 200 ms*). Selecting $I_{max} = 1.1$ pu is intended to confirm that the *Remote* system is "weaker", as compared to Case 1 when $I_{max} = 1.2$ pu.

Voltage, current, and active and reactive power are shown in Figure 9.16 for the GFMI at Aghada, the fault location. Unstable dynamics can be seen for *VI – 305 ms*, relative to the 260 ms fault duration scenario, while Figure 9.16b confirms that the inverter current is limited within 1.1 pu for both VI and current scaling, although the current is transiently higher with VI control (affecting converter lifetime). Figure 9.16c and d shows that the GFMI is stable for the 260 ms fault, although it is comparatively slow ($\approx 1.3\,\text{s}$) to recover. The inverter goes unstable for the 305 ms fault with VI control once the fault is cleared, since, for an extended period ($>1.8\,\text{s}$), the voltage and reactive power are low, and, at the same time, the active power falls highly negative (-1 pu), before returning to their pre-fault values. The instability is due to the extended acceleration of the inverter angular speed, with the regulation capability also lost, given that the injected current is at its limit. Later it will be seen that virtual angular speed freezing extends the grid-forming fault critical clearing time.

Current scaling also results in the inverter being unstable for a 200 ms fault, as the post-fault voltage is low and the active power falls highly negative, Figure 9.16a and b. Similar to Cases 1 and 2, the stable *VI – 260 ms* and unstable *Scaling – 200 ms* scenarios again indicate that soft current limiting control leads to more robust grid-forming transient stability than hard saturation limiting.

Now considering the converter voltage, Figure 9.16a indicates that the voltage stays low for an extended period ($>1.3\,\text{s}$) for the two VI scenarios, while even for the stable *VI – 260 ms* scenario, voltage recovery is slow ($\approx 1.3\,\text{s}$), suggesting that the fault critical clearing time at the Aghada location is close to 260 ms using VI current limit control. Comparing *VI – 260 ms* here ($I_{max}^{VI} = 1.1$ pu) against that in Case 1 ($I_{max}^{VI} = 1.2$ pu) indicates that the system takes longer to recover, noting the lower current limit.

9.3.3.2 GFMI Transient Stability Enhanced by Virtual Angular Speed Freezing

Case 4: Previously in Cases 1–3, VI control was seen to be the preferred option, but stability was affected for faults longer than ≈ 260 ms. Freezing of the virtual angular speed (Equation 9.10) is now introduced for the same fault conditions as Case 3: (a) VI control combined with virtual angular speed freezing for a 505 ms, three-phase bolted fault at Aghada (*VI + Frozen – 505 ms*); (b) *Scaling + Frozen – 505 ms*; and (c)

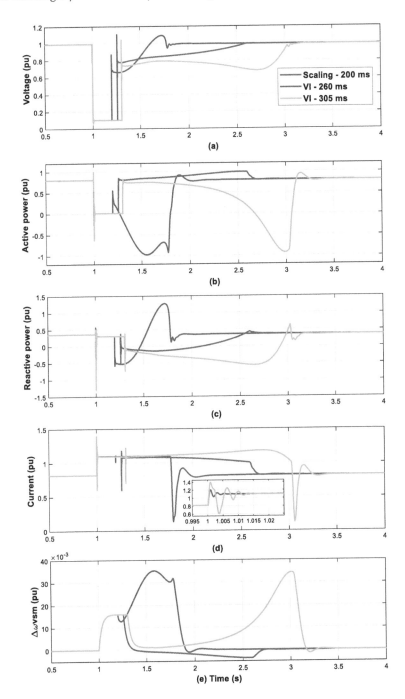

FIGURE 9.16 Impact of virtual impedance and scaling current reference saturation controls, and fault duration on the grid-forming transient stability at Aghada ((a), (b), (c), (d) and (e), respectively) for 200, 260, and 305 ms three-phase bolted faults at the Aghada bus for the *Remote* Irish grid with an overcurrent limit of 1.1 pu under Case 3.

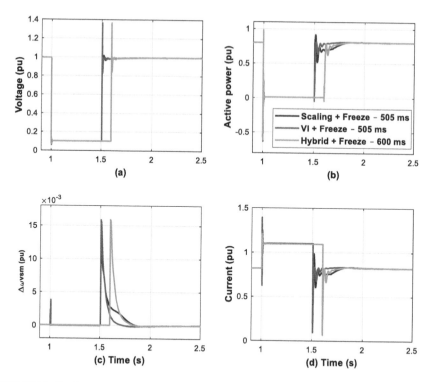

FIGURE 9.17 Impact of virtual impedance and scaling current reference saturation controls, and freezing of virtual angular speed on the grid-forming inverter at Aghada ((a), (b), (c) and (d), respectively) for 505 and 600 ms three-phase bolted faults at the Aghada bus of the *Remote* Irish grid with an overcurrent limit of 1.1 pu (Case 4).

both current limiting control approaches for a 600 ms fault (*Hybrid + Freeze – 600* ms). In order to achieve a fair comparison with Case 3, the current limiting values I_{max}^{VI}, I_{max} for (a) and (b) are unchanged at 1.1 pu, and for (c), 1.1 and 1.15 pu, respectively, for VI and scaling.

Results for the GFMI at Aghada are shown in Figure 9.17, and, as previously, the "steady-state" current is limited within 1.1 pu for both the VI and current scaling approaches, Figure 9.17d. The initial post-fault voltage for *VI + Frozen – 505* ms, Figure 9.17a, is close to the pre-fault value, and, significantly, 0.2 pu higher when compared against *VI – 260* ms, Figure 9.16a. The voltage and active power also recover much more quickly ($\approx$0.1 s to reach the pre-fault state, relative to $\approx$1.3 s). It can be seen in Figure 9.17c that freezing the virtual angular speed increases the critical clearing time for the *VI, Scaling* and *Hybrid* approaches, with inverter stability maintained, even for 600 ms faults. Figure 9.17b shows that compared to the other VI scenarios, the scaling approach results in slightly lower post-fault active power, but the inverter quickly recovers to its pre-fault state. In conclusion, freezing the virtual angular speed improves the transient stability of the GFMI, irrespective of the current limiting strategy, with an improved post-fault initial voltage, a faster post-fault recovery, and an extended fault ride through capability.

9.4 CONCLUSIONS

The transient stability of a power system containing a high share of GFMIs can be significantly different to that of a traditional synchronous generator-based system, recognising the lower capacity (current) headroom, but flexible control capability, of power electronic inverters. System stability was investigated for a future Irish grid consisting entirely of GFMIs under three-phase fault conditions with the inverters placed at existing locations for large-scale conventional generation. EMT simulations showed that a 100% GFMI system, employing either droop control (or VSM), dVOC control, or a mix of both, under a combination of VI and scaling current saturation limiting control, was robust against three-phase faults, with consistent performance being achieved, despite variations in fault location or inverter control methods.

A modified grid configuration was also investigated, with the inverters being placed at more remote locations, recognising likely locations for wind farms. Simulation results show that the *Remote* system is weaker than the above *Urban* system. For a reduced overcurrent limit, the GFMIs typically took more than 1 s to recover to their pre-fault state under VI current limiting control, when 260 ms, three-phase faults were applied, and they could go unstable with the current scaling saturation approach for a 200 ms fault. However, by freezing the virtual angular speed during the fault, for both the VI and current scaling approaches, the GFMIs remained stable, even with a reduced current limit, and the pre-fault state was quickly recovered, even for 600 ms faults, indicating an increased fault critical clearing time. Time domain simulations also show that when active or reactive current prioritisation current saturation controls are applied that GFMIs can introduce large, high-frequency resonance oscillations, but a scaling-down current saturation approach can help to mitigate such problems by generating smoother current references.

REFERENCES

[1] RTE, "Description of system needs and test cases," MIGRATE WP3, 2016. Available: https://www.h2020-migrate.eu/downloads.html [Accessed: 27/5/21].

[2] Hodge, B.-M., Jain, H., Brancucci, C., et al., "Addressing technical challenges in 100% variable inverter-based renewable energy power systems," *Wiley Interdiscip. Rev. Energy Environ.*, 9(5), p. e376, 2020.

[3] IEEE PES, "Stability definitions and characterization of dynamic behavior in systems with high penetration of power electronic interfaced technologies," Tech. Rep. PES-TR77, 2020.

[4] Holttinen, H., Kiviluoma, J., Flynn, D., et al., "System impact studies for near 100% renewable energy systems dominated by inverter based variable generation," *IEEE Trans. Power Syst.*, 37(4), pp. 3249–3258, 2022.

[5] Zhao, X., and Flynn, D., "Transient stability enhancement with high shares of grid-following converters in a 100% converter grid," in *IEEE PES Innovative Smart Grid Technologies Europe (ISGT-Europe)*, pp. 594–598, 2020. A virtual event from 26 to 28 October 2020.

[6] Matevosyan, J., Badrzadeh, B., Prevost, T., et al., "Grid-forming inverters: Are they the key for high renewable penetration?," *IEEE Power Energy Mag.*, 17(6), pp. 89–98, 2019.

[7] Chandorkar, M.C., Divan, D.M. and Adapa, R., "Control of parallel connected inverters in standalone AC supply systems," *IEEE Trans. Ind. Appl.*, 29(1), pp. 136–143, 1993.

[8] Zhang, L., Harnefors, L. and Nee, H.P., "Power-synchronization control of grid-connected voltage-source converters," *IEEE Trans. Power Syst.*, 25(2), pp. 809–820, 2010.

[9] Simpson-Porco, J.W., Dörfler, F. and Bullo, F., "Synchronization and power sharing for droop-controlled inverters in islanded microgrids," *Automatica*, 49(9), pp. 2603–2611, 2013.

[10] Zhong, Q.C. and Weiss, G., "Synchronverters: Inverters that mimic synchronous generators," *IEEE Trans. Ind. Electron.*, 58(4), pp. 1259–1267, 2010.

[11] D'Arco, S., Suul, J.A. and Fosso, O.B., "A virtual synchronous machine implementation for distributed control of power converters in smart grids," *Electr. Power Syst. Res.*, 122, pp. 180–197, 2015.

[12] Guan, M., Pan, W., Zhang, J., et al., "Synchronous generator emulation control strategy for voltage source converter (VSC) stations," *IEEE Trans Power Syst.*, 30(6), pp. 3093–3101, 2015.

[13] Cvetkovic, I., Boroyevich, D., Burgos, et al., "Modeling and control of grid-connected voltage-source converters emulating isotropic and anisotropic synchronous machines," in *IEEE Workshop on Control Model. Power Electron. (COMPEL)*, 2015.

[14] Arghir, C., Jouini, T. and Dörfler, F., "Grid-forming control for power converters based on matching of synchronous machines," *Automatica*, 95, pp. 273–282, 2018.

[15] Huang, L., Xin, H., Wang, Z., et al., "A virtual synchronous control for voltage-source converters utilizing dynamics of DC-link capacitor to realize self-synchronization," *IEEE J. Emerg. Sel. Topics Power Electron.*, 5(4), pp. 1565–1577, 2017.

[16] B. Johnson, M. Sinha, N. Ainsworth, et al., "Synthesizing virtual oscillators to control islanded inverters," *IEEE Trans. Power Electron.*, 31(8), pp. 6002–6015, 2015.

[17] Seo, G.-S., Colombino, M., Subotic, I., et al., "Dispatchable virtual oscillator control for decentralized inverter-dominated power systems: Analysis and experiments," in *IEEE Appl. Power Electron. Conf. Expo.*, pp. 561–566, 2019. From 17 to 20 June 2019, Toronto, Ontario, Canada.

[18] Musca, R., Vasile, A. and Zizzo, G., "Grid-forming converters. a critical review of pilot projects and demonstrators," *Renew. Sust. Energ. Rev.*, 165, p. 112551, 2022.

[19] Gevorgian, V., Shah, S., Yan, W., et al., "Grid-forming wind power," ESIG Spring Technical Workshop, Tucson, AZ, 2022. https://www.nrel.gov/docs/fy22osti/82509.pdf.

[20] Milano, F., Dörfler, F., Hug, et al., "Foundations and challenges of low-inertia systems," in *Proc. Power Syst. Comput. Conf.*, Ireland, 2018.

[21] Sadeghkhani, I., Golshan, M.E.H., Guerrero, et al., "A current limiting strategy to improve fault ride-through of inverter interfaced autonomous microgrids," *IEEE Trans. Smart Grid*, 8(5), pp. 2138–2148, 2016.

[22] Huang, L., Xin, H., Wang, et al., "Transient stability analysis and control design of droop-controlled voltage source converters considering current limitation," *IEEE Trans. Smart Grid*, 10(1), pp. 578–591, 2017.

[23] Taul, M.G., Wang, X., Davari, P. et al., "Current limiting control with enhanced dynamics of grid-forming converters during fault conditions," *IEEE J. Emerg. Sel. Topics Power Electron.*, 8(2), pp. 1062–1073, 2019.

[24] Paquette, A. D., and Divan, D. M., "Virtual impedance current limiting for inverters in microgrids with synchronous generators," *IEEE Trans. Ind. Appl.*, 51(2), pp. 1630–1638, 2014.

[25] Groß, D. and Dörfler, F., "Projected grid-forming control for current-limiting of power converters," In *Allerton Conference on Communication, Control, and Computing*, pp. 326–333, 2019. From 24 to 27 Sep 2019, Monticello, Illinois, USA.

[26] Zhao, X. and Flynn, D., "Grid-forming converter current limiting design to enhance transient stability for grid phase jump events," in *11th IFAC Symposium on Control of Power and Energy Systems (CPES)*, 2022. Virtual Conference from 21 to 23 June, 2022.

[27] Poolla, B.K., Groß, D. and Dörfler, F., "Placement and implementation of grid-forming and grid-following virtual inertia and fast frequency response," *IEEE Trans. Power Syst.*, 34(4), pp. 3035–3046, 2019.

[28] Yang, C., Huang, L., Xin, H. et al., "Placing grid-forming converters to enhance small signal stability of PLL-integrated power systems," *IEEE Trans. Power Syst.*, 36(4), pp. 3563–3573, 2020.

[29] Thakurta, P., Zhao, X., Flynn, D., "Local control and simulation tools for large transmission systems," EU H2020 MIGRATE WP3, 2019.

[30] Zhao, X., Thakurta, P.G. and Flynn, D., "Grid-forming requirements based on stability assessment for 100% converter-based Irish power system," *IET Renew. Power Gener.*, 16(3), pp. 447–458, 2022.

[31] Qoria, T., Cossart, Q., Li, C., Guillaud, et al., "Deliverable 3.2: Local control and simulation tools for large transmission systems," MIGRATE project, 2018.

[32] Zhao, X. and Flynn, D., "Freezing grid-forming converter virtual angular speed to enhance transient stability under current reference limiting," in *IEEE 21st Workshop on Control and Modeling for Power Electronics (COMPEL 2020)*, 2020. Virtual Conference from 9 to 12 November, 2020.

[33] Zhao, X. and Flynn, D., "Stability enhancement strategies for a 100% grid-forming and grid-following converter-based Irish power system," *IET Renew. Power Gener.*, 16(1), pp. 125–138, 2022.

[34] Rokrok, E., Qoria, T., Bruyere, et al., "Transient stability assessment and enhancement of grid-forming converters embedding current reference saturation as current limiting strategy," *IEEE Trans. Power Syst.*, 2021. DOI: 10.1109/TPWRS.2021.3107959.

[35] Groß, D., Colombino, M., Brouillon, J.-S., et al., "The effect of transmission-line dynamics on grid-forming dispatchable virtual oscillator control," *IEEE Trans. Contr. Netw. Syst.*, 6(3), pp. 1148–1160, 2019.

[36] Seo, G.S., Colombino, M., Subotic, I., et al., "Dispatchable virtual oscillator control for decentralized inverter-dominated power systems: Analysis and experiments," in *Proc. IEEE Appl. Power Electron. Conf. Expo. (APEC)*, 2019. From 17 to 22 March, 2019, Anaheim, California, USA.

[37] Dymola Systems Engineering, https://www.3ds.com/products-services/catia/products/dymola/.

[38] Zhao, X. and Flynn, D., "Grid-forming converter angular speed freezing to enhance transient stability in 100% grid-forming and mixed power systems", in *11th IFAC Symposium on Control of Power and Energy Systems (CPES)*, 2022. Virtual Conference from 21 to 23 June, 2022.

[39] EirGrid, "All-Island ten year transmission forecast statement 2020," 2020.

[40] EirGrid and SONI, "Shaping our electricity future," Technical Report, 2021.

[41] EirGrid and SONI, "Operational Constraints Update 28/06/2021," 2021.

[42] EirGrid, "Tomorrow's energy scenarios consultation Ireland - Planning our energy future," 2019.

[43] Zhao, X. and Flynn, D., "Stability enhancement strategies for a 100% grid-forming and grid-following converter-based Irish power system," *IET Renew. Power Gener.*, 16(1), pp. 125–138, 2022.

10 Virtual Oscillator-Controlled Grid-Forming Inverters Incorporating Online Parametric Grid Impedance Identification

Nabil Mohammed and Behrooz Bahrani
Monash University

Mihai Ciobotaru
Macquarie University

CONTENTS

DOI: 10.1201/9781003302520-10

10.1 OVERVIEW

Dispatchable virtual oscillator control (dVOC) is an emerging communication-free control strategy for grid-forming (GFM) inverters. In contrast to droop control which operates based on phasor quantities, dVOC is a nonlinear and time-domain approach and stabilizes arbitrary initial conditions to a sinusoidal steady state. It has been recently emphasized that the performance (e.g., stability) of dVOC-based inverters is affected by the literature grid impedance seen by the inverter at its point of common coupling (PCC). Therefore, it becomes necessary to enable these inverters to online estimate the grid impedance, which has not yet been explored in the literature. Therefore, this chapter proposes the implementation of an online grid impedance estimation into the control loop of a single-phase dVOC-based inverter operated in a grid-connected mode. The identification of the grid impedance is achieved in two stages. First, the pseudorandom binary sequence (PRBS) disturbance is injected on the top of the voltage reference provided to the inverter's pulse width modulation (PWM), and the broadband grid impedance is estimated online based on the PCC current and voltage response. Second, the parametric resistive-inductive (RL) model of the grid impedance is estimated by curve fitting of the impedance obtained in the first stage. The presented approach to measure the grid impedance is software-based and requires only local voltage and current measurements; hence, it is cost-effective and practical. Simulation and hardware-in-the-loop (HIL) experimental results validate the high accuracy of the proposed approach to estimate the grid impedance, which can be used for impedance-related applications to dVOC-based inverters (e.g., real-time stability analysis).

10.2 INTRODUCTION

With the increasing penetration level of distributed energy resources (DERs) into the power systems, GFM inverters are becoming an inevitable part of future AC power systems. Originally, GFM inverters are proposed for micro and islanded grid applications. Nowadays, GFM inverters are considered the key enabling technology for grid-connected applications as they have the potential to effectively enhance the system stability and resiliency, enable high penetration of power electronics-based generation, and strengthen the weak parts of the grid [1].

GFM inverters can be operated in both grid-connected mode and in islanded mode. In the grid-connected mode, GFM inverters behave as controlled current sources, where the main goal is to control the injected active and reactive power to the grid. On the other hand, in the islanded mode, GFM inverters behave as controlled voltage sources that directly regulate the PCC voltage and frequency [2].

Several techniques have been proposed to control GFM inverters, including droop control, synchronous-machine-based control, and virtual oscillator control (VOC) [1]. In the literature, droop control is widely used [2]. However, since VOC is shown to outperform droop control, VOC is gaining increased popularity [3–5], with VOC becoming an emerging control technique for GFM inverters [6–15]. It mimics the dynamics of weakly nonlinear Van der Pol oscillators to control the inverter using only the output current measurement. VOC is a time-domain control that is characterized by its rapid response and its ability to stabilize [3–5].

VOC is originally proposed for controlling a single-phase inverter operating in the islanded mode [6], and it is extended for three-phase AC islanded microgrids [7]. The conditions for synchronization of parallel single- and three-phase inverters are also presented in references [6,7]. The design requirements/constraints and the synthesis of VOC are addressed in detail in reference [8]. Nevertheless, the original VOC approach [6,7] is not dispatchable as power commands cannot be specified in the VOC control loop. Therefore, for operation in the grid-connected mode, dispatchable VOC (dVOC) is proposed to allow the regulation of active and reactive power.

A dVOC is developed for three-phase inverters in reference [9]. A dVOC relying on a complex-valued parameter connected to a non-stiff grid is presented in reference [11]. A unified voltage oscillator controller (uVOC) is proposed to enable a unified analysis, design, and implementation for both GFM and grid-following (GFL) inverters in reference [12]. By contrast, as single-phase dVOC-based inverters have several potential applications, including grid voltage and frequency support to enhance the integration of small-scale renewable energy sources, dVOC is developed also for single-phase inverters. A dVOC based on active and reactive power reference tracking strategies in microgrid applications is proposed in references [14,15]. Additionally, the work in reference [9] has been experimentally verified for single-phase inverters in reference [10]. However, the presented methods in references [9,14,15] lack a suppression scheme for the third harmonic contents in the inverter output current, limiting the application of these techniques as the output current harmonics are defined by power quality standards. A hierarchical control strategy is proposed to allow the operation in both grid-connected mode and islanded mode in reference [13]. However, in the grid-connected mode, the slow response of the control structure is the main drawback of such single-phase VOC-based systems due to the use of three control layers (primary, secondary, and tertiary).

Furthermore, the equivalent grid impedance seen at the PCC can affect the operation of dVOC-based inverters [16,17]. The effects of the transmission-line dynamics on the stability analysis of dVOC GFM inverters are discussed in reference [16]. The line impedance matrix seen by the VOC-based inverter at its terminal is used in the analysis. However, obtaining the information of the line impedance is not discussed. In reference [17], it was shown how the resistive grid negatively impacts the power reference tracking control of Andronov-Hopf Oscillator Control-based inverters. Then, a modification based on the grid impedance information was proposed to mitigate these errors and enable simultaneous control of both active and reactive power. However, obtaining the information of the grid impedance has not been discussed.

In summary, there is a need to further investigate the implementation of an online grid impedance estimation algorithm into the control of VOC-based inverters. This approach is a cost-effective approach, where extra hardware is not needed. This is important as the equivalent grid impedance is a time-varying parameter. Previous studies on other control techniques for grid-connected inverters have shown that several online grid impedance estimation techniques can be embedded into control loops of inverters [18–23], where the estimated impedance information has been used as a useful tool for several applications to enhance the operation of the inverters [24–28]. Therefore, similar approaches can be employed for VOC-based inverters to ensure accurate estimation results for the grid impedance components, where the obtained online information of grid impedance can be adopted to enhance the performance of the dVOC inverters as illustrated in reference [17]. To the best of the authors' knowledge, there is no implementation of online grid impedance estimation techniques for either VOC-based or dVOC-based inverters in the literature.

This chapter proposes the implementation of online grid impedance estimation for dVOC-based inverters. The estimation procedure consists of two stages: the wideband estimation, followed by the parametric estimation of the RL model of the grid impedance. The contributions of this chapter can be summarized as:

1. To enable online broadband estimation of the grid impedance, the grid impedance estimation based on PRBS injection is incorporated in the control loop of the dVOC inverter.
2. To obtain the parametric RL model of the grid impedance, a complex curve fitting of the broadband impedance is presented for dVOC inverters.

The rest of the chapter is structured as follows. Section 10.2 provides a review on the conventional VOC used for power inverters. Section 10.3 presents in detail the dispatchable power control for VOC-based inverters, and the online broadband and parametric estimation of the grid impedance. Sections 10.4 and 10.5 provide simulation and HIL experimental results, respectively, to evaluate the performance of the proposed methodology. Lastly, Conclusions are discussed in Section 10.6.

10.3   CONVENTIONAL VOC

Figure 10.1 illustrates the implementation of the conventional Van der Pol virtual oscillator controller for a single-phase inverter system. The inverter is connected with the rest of the system through an LCL filter to attenuate high-order switching harmonics in the inverter terminal voltage. v_{dc} and i symbols denote the input dc-link voltage and the inverter output current, respectively.

The virtual oscillator is coupled to the physical electrical system through voltage- and current-scaling factors that are denoted by k_v and k_i, respectively. The input of the virtual oscillator is the measured output current i, scaled by k_i. The output of the virtual oscillator is the virtual capacitor voltage v_c, which is scaled by k_v to produce v. Then, v is used to generate the PWM signals required to drive the inverter. If V is the rms voltage and $\varphi(t)$ is the instantaneous phase, v is given by

$$v = \sqrt{2}V\cos(\varphi).$$ (10.1)

The Van der Pol virtual oscillator consists of the parallel connection of (a) a cubic voltage-dependent current source αv_c^3, where α is the cubic-current source constant,

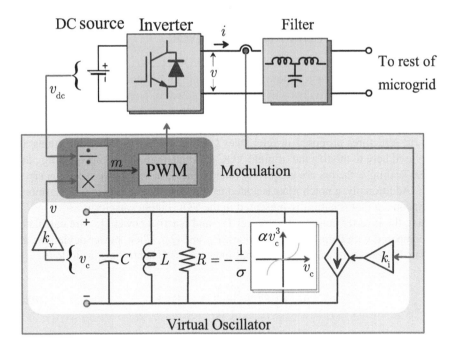

FIGURE 10.1 Implementation of the conventional virtual oscillator controller for a single-phase power inverter.

(b) a negative-conductance $-\sigma$, and (c) a harmonic oscillator composed of virtual inductance L and virtual capacitance C with a resonant frequency $\omega^* = 1/\sqrt{LC}$. ω^* should be equal to the nominal frequency of the electric system (e.g., $2\pi 50$ or $2\pi 60$) [8].

The dynamics of the system are given by the following nonlinear differential equations:

$$L\frac{di_L}{dt} = \frac{v}{k_v},\tag{10.2}$$

$$C\frac{dv}{dt} = -\alpha\frac{v^3}{k_v^2} + \sigma v - k_v i_L + k_v k_i i,\tag{10.3}$$

where i_L is the current of the virtual inductor.

In the islanded mode, it is shown that the VOC-based inverters exhibit droop-like behavior [3,4]. Hence, the frequency and voltage of the VOC-based inverters deviate from the nominal values depending on both the real and reactive power drawn [13]. The deviation problem is of more concern in the grid-connected operation of the system as the main interest here is to accurately control the dispatched active and reactive power by the inverter. Therefore, dVOC has been used for power dispatch, as illustrated in the next section.

10.4 PROPOSED IMPLEMENTATION OF ONLINE GRID IMPEDANCE ESTIMATION FOR dVOC

Figure 10.2 shows the control structure of the dVOC operated in the grid-connected mode. The active and reactive power deviations are eliminated by an external power control loop. Hence, zero errors in tracking the power references/commands can be achieved.

It is worth mentioning that the dVOC shown in Figure 10.2 is slightly different from the structures proposed in references [14,15], where the power deviations are being used here to modify the original VOC in which both the voltage-scaling $_v$ and current-scaling k_i factors are adaptively tuned to achieve zero power tracking errors rather. Additionally, a notch filter is added to suppress the dominant third harmonic ($h = 3$) content that is naturally presented in the VOC output waveform.

First, the inverter output active power P_{pcc} and reactive power Q_{pcc} are calculated and compared with their command values, P_{ref} and Q_{ref}. Then, power deviations are regulated toward zero using proportional-integrator (PI) controllers, as follows:

$$\delta k_v = \left(K_p^p + \frac{K_i^p}{s} \right)\left(P_{ref} - P_{pcc} \right), \tag{10.4}$$

$$\delta k_i = \left(K_p^q + \frac{K_i^q}{s} \right)\left(Q_{ref} - Q_{pcc} \right), \tag{10.5}$$

where K_p^p, K_i^p, K_p^q, and K_i^q are the parameters of power controllers. After that, the processed errors in the active power (δk_v) and reactive power (δk_i) are sent to the VOC control loop. The control laws of the modified dVOC to achieve zero tracking errors in both active and reactive power dispatch are summarized as

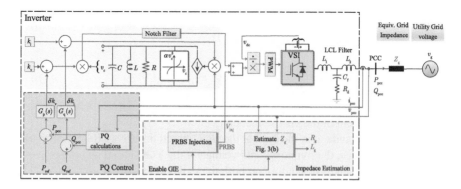

FIGURE 10.2 Dispatchable virtual oscillator control for a single-phase grid-connected inverter integrating online parametric broadband grid impedance estimation.

$$k_v = k_{v_0} + \delta k_v, \qquad (10.6)$$

$$k_i = k_{i_0} - \delta k_i. \qquad (10.7)$$

As shown in Figure 10.2, estimation of the grid impedance is performed during the normal operation of the dVOC-based inverter, and it requires only the local PCC voltage and current measurements. Hence, it is reliable as it does not employ communication links or extra hardware.

As shown in Figures 10.2 and 10.3, the implemented methodology to achieve accurate grid impedance identification requires only the local PCC voltage and current measurements. Hence, it is reliable as it does not employ communication links or extra hardware. The estimation consists of two steps. First, the PRBS disturbance is injected into the grid by the dVOC-based inverter. Then, the wideband non-parametric grid impedance is estimated based on the discrete Fourier transformation (DFT). Second, the parametric grid impedance estimation is obtained based on the curve fitting. These two steps are presented in the following two subsections, respectively.

10.5 NON-PARAMETRIC GRID IMPEDANCE ESTIMATION BASED ON PRBS INJECTION

10.5.1 DISTURBANCE INJECTION

PRBS is employed in this chapter to estimate the grid impedance over the frequency range of interest. The parameters used for the PRBS should be chosen carefully to obtain an accurate impedance estimation. These parameters are bit-length of the PRBS generator, the frequency resolution (F_{res}), and the injection amplitude (V_{inj}). Additionally, the impedance estimation using power converters is limited to several constraints including the switching and sampling frequencies. Among PRBS types, the maximum length binary sequence (MLBS) is chosen in this study due to its simple implementation. A 12-bit MLBS is used with a frequency resolution of 1 Hz. Hence, the required time to inject a full period of PRBS into the grid is 1 s. Further details on the PRBS parameters are presented in the results section.

The details of the PRBS injection are presented in Figures 10.2 and 10.3 in which the dVOC-based grid-connected inverter is used to inject the PRBS into the grid. Once the enabling signal for grid impedance estimation (GIE) is received, the PRBS is added to the inverter voltage reference. The injection lasts for at least two full PRBS injection periods. Simultaneously with the PRBS injection, the PCC voltage and current are collected to be processed for the impedance estimation as will be explained in the next subsection.

10.5.2 NON-PARAMETRIC GIE

Once the grid-connected inverter starts injecting the PRBS into the grid, the GIE unit is activated using the same enabling signal that is used for PRBS injection.

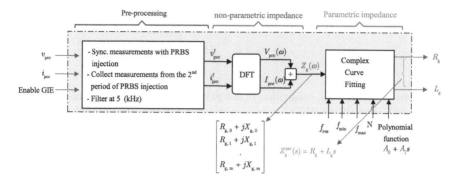

FIGURE 10.3 Detailed implementation of the online non-parametric and parametric estimation of the grid impedance using PRBS.

Details on the estimation procedure of the grid impedance are shown in Figure 10.3. First, PCC voltage and current are measured using the existing current and voltage sensors. Then, these measurements are pre-processed (e.g., filtering). To remove the PRBS transient, two PRBS periods are injected into the system where only the PCC voltage and current measurements of the second-period injection are used. After that, the grid impedance vector is estimated in the frequency domain for the voltage and current measurements in the time domain by means of DFT, as follows:

$$Z_g(\omega) = \frac{V_{pcc(\omega)}}{I_{pcc(\omega)}} (\Omega), \tag{10.8}$$

where $V_{pcc(\omega)}$ and $I_{pcc(\omega)}$ are the voltage and current measurements in the frequency domain.

10.6 PARAMETRIC IMPEDANCE BASED ON COMPLEX CURVE FITTING

Estimating the grid impedance using PRBS has several constraints. First, the obtained grid impedance spectrum is limited up to half of the switching frequency (as per Shannon Sampling Theorem). Second, the PRBS estimation technique provides very accurate results up to a third of its generation/clock frequency f_{gen}. This constraint is due to the drop in the PRBS power spectrum density (PSD) to more than 3 dB after $f_{gen}/3$. According to reference [29], the maximum bandwidth of interest $\left(f_{BW-max}\right)$ is

$$f_{BW-max} = \frac{1}{2.5} \times f_{gen} (Hz). \tag{10.9}$$

When obtaining an accurate continuous model of grid impedance for all the frequency ranges of interest, a good solution is to construct the mathematical transfer function of the grid impedance by adopting a curve fitting method. The curve fitting for parametric grid impedance purposes is performed through a chosen series of data points. This chapter adopts a complex curve fitting technique [30] to achieve this purpose. Its working principle is as follows: "It is based on the minimization of the weighted sum of the squares of the errors between the absolute magnitudes of the actual function and the polynomial ratio, taken at various values of frequency (the independent variable). The problem of the evaluation of the unknown coefficients is reduced to that of the numerical solution of certain determinants. The elements of these determinants are functions of the amplitude ratio and phase shift, taken at various values of frequency" [30].

Figure 10.3 shows the inputs, initialization parameters, and outputs of the curve fitting algorithm. The initial output of the algorithm is the non-parametric grid impedance vector obtained (10.8). The required initializations to obtain the parametric grid impedance are the PRBS frequency resolution (f_{res}), frequency range of the data points to be fitted [$f_{min} - f_{max}$], the selected number of the data points (N) to be processed for the curve fitting, and polynomial function for the curve fitting. Further details on the effects of these parameters on the curve fitting results can be found in reference [22]. Since the actual grid impedance transfer function used in this chapter is the RL model, the best candidate polynomial function for curve fitting ($Z_g^{par}(s)$) is given by

$$Z_g^{par}(s) = A_0 + A_1 s, \tag{10.10}$$

where A_0 and A_1 are the transfer function coefficients that need to be estimated. Lastly, the R and L components of the grid impedance are obtained by comparing the RL model with equation 10.10 as

$$\begin{cases} R_g = A_0, \\ L_g = A_1. \end{cases} \tag{10.11}$$

10.7 SIMULATION RESULTS

To validate the accuracy of the impedance estimation, the single-phase dVOC-based inverter shown in Figure 10.3 is simulated in MATLAB/Simulink software. The power stage and control parameters used for the simulation are listed in Table 10.1. Moreover, Table 10.2 summarizes both the parameters used for PRBS in this chapter and the parameters used for the curve fitting algorithm in order to obtain accurate impedance estimation results.

Three different scenarios are investigated. These case studies are to test the performances of the (a) the active and reactive power dispatch, (b) the online wideband and parametric GIE, and finally (c) the effects of implementing the notch filter in the control loop on suppressing the third harmonic in the inverter output current.

TABLE 10.1
System and Control Parameters

Quantity	Symbol	Value	Unit
Grid parameters			
Nominal voltage (RMS)	V_g^*	230	V
Nominal frequency	f_g^*	50	Hz
Grid impedance	R_g	0.5	Ω
	L_g	0.2	mH
R/X ratio of grid impedance	R/X	7.957	Ω/Ω
Inverter power stage parameters			
Inverter nominal active power	P_n	5	kW
Inverter nominal reactive power	Q_n	5	kVar
Filter inverter-side inductance	L_1	2.5	mH
Filter grid-side inductance	L_2	0.9748	mH
Filter capacitance	C_f	4.7	uF
Damping resistance	R_d	3.3	Ω
Inverter VOC control parameters			
Current feedback gain	$k_{i\,0}$	0.0432	A/A
Voltage scaling factor	$k_{v\,0}$	253	V/V
Conductance	σ	4.3381	Ω^{-1}
Cubic-current source coefficient	α	2.8921	A/V3
Oscillator inductance	L	54.415	uH
Oscillator capacitance	C	0.1862	F
Switching frequency	f_{sw}	10	kHz
Power control			
Active power gains	K_p^p	2.5×10^{-3}	–
	K_i^p	40×10^{-3}	1/sec
Reactive power gains	K_p^q	50×10^{-6}	–
	K_i^q	300×10^{-6}	1/sec

TABLE 10.2
Parameters of the PRBS and of Grid Impedance Curve Fitting Algorithm

Description	Symbol	Value	Unit
PRBS			
Bit-length of the PRBS generator	–	$12 - bit$	–
Maximum length binary sequence	MLBS	4,095	bits
Frequency resolution	f_{res}	1	Hz
Injected disturbance amplitude	V_{inj}	10 (4.3% V_g^*)	V
The curve fitting algorithm			
Minimum frequency	f_{min}	5	Hz
Maximum frequency	f_{max}	2,000	Hz
No. of the impedance data points used for curve fitting	N	45	–

10.7.1 ACTIVE AND REACTIVE POWER DISPATCH

Figure 10.4 presents the output active and reactive power track the reference power command only after enabling the proposed control at 4 s. By neglecting the power transients, it can be observed that the proposed control has zero steady-state errors in both the active and reactive power.

Figure 10.5 shows the waveforms of output current that correspond to the active and reactive power commands.

As stated earlier, the power control is accomplished through the adaptive tuning of k_v and k_i parameters of the dVOC-based inverter. Figure 10.6 presents these parameters during the various simulated scenarios of the active and reactive power commands.

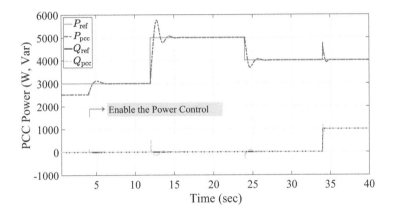

FIGURE 10.4   Simulation waveforms of output active and reactive power before and after enabling the power control.

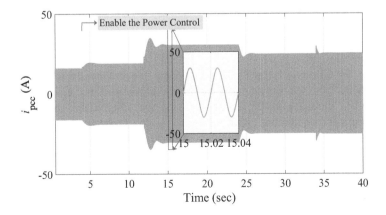

FIGURE 10.5   Simulation waveforms of the output current before and after enabling the power control.

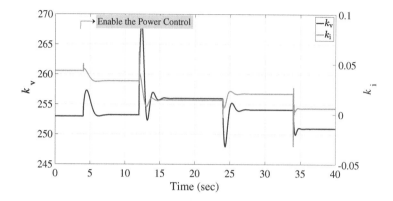

FIGURE 10.6 Simulation waveforms of adaptive tuning of the k_v and k_i parameters of the dVOC-based inverter.

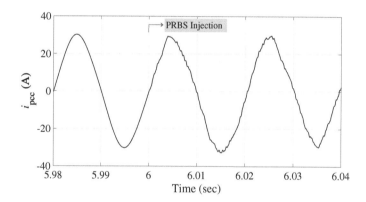

FIGURE 10.7 Simulation waveforms of the output current before and under PRBS injection for online broadband grid impedance estimation.

10.7.2 ONLINE GIE

To achieve the GIE, two full periods of PRBS are injected into the system. Only the measurements from the second one are to be used for impedance estimation.

Figure 10.7 shows the waveforms of output current. The PRBS injection of the two periods starts at 6 s and it has lasted for 2 s.

Figure 10.8 gives the estimated non-parametric grid impedance with a frequency resolution of 1 Hz; see Table 10.2. The selected data points ($N=45$ points) are also shown in the same figure. Figure 10.9 compares the obtained parametric impedance from curve fitting-based equation (10.10) with the reference value of the grid impedance. The estimation errors in the grid impedance components are between the identified parametric model ($Z_{g\,par} = 0.5025 + 0.0001993\ s$) and the reference model ($Z_{g\,ref} = 0.5 + 0.0002\ s$) are $\Delta R_g\% = 0.49$ and $\Delta L_g\% = -0.35$.

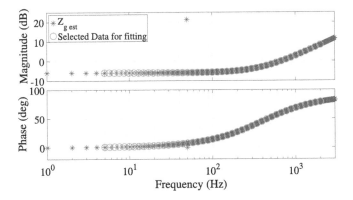

FIGURE 10.8 Non-parametric estimation of the grid impedance.

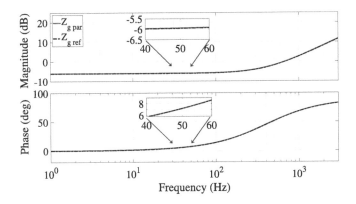

FIGURE 10.9 Parametric estimation of the grid impedance.

10.7.3 Effects of Notch Filter on THD%

To verify the performance of the notch filter, simulation results are conducted in this section with and without applying a notch filter that is tuned to suppress the dominant third harmonic ($h=3$) content naturally present in the VOC output waveform. Figures 10.10 and 10.11 compare the frequency spectrum of the PCC current without and with notch filter, respectively. It can be observed that the third harmonic, located at 150 Hz, is significantly reduced after adding the notch filter to the control loop of the inverter.

10.8 HIL EXPERIMENTAL RESULTS

The single-phase dVOC-based grid-connected inverter shown in Figure 10.2, whose power stage and control parameters are listed in Tables 10.1 and 10.2, is investigated in the laboratory. The HIL experimental test rig is presented in Figures 10.12 and

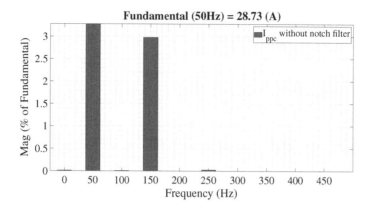

FIGURE 10.10   Simulation results of the spectral contents of the currents injected into the grid without notch filter.

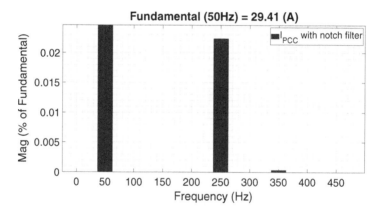

FIGURE 10.11   Simulation results of the spectral contents of the currents injected into the grid with notch filter.

10.13. The HIL experiments are executed in the PLECS RT Box real-time simulation system. The control algorithm is realized through a TMS320F28069M LaunchPad from Texas Instruments.

In this section, two different cases have been studied. In Case 1, the performance of the dispatchable power control is tested first for different active and reactive power references. In Case 2, the online broadband and parametric estimation of the grid impedance is investigated.

10.8.1   ACTIVE AND REACTIVE POWER DISPATCH

Figure 10.14 shows the dynamic response of the dVOC-based inverter in grid-connected mode to different power references. Initially, the power references are set to $P_{ref} = 3\,\text{kW}$ and $Q_{ref} = 0$ kVar. However, the measured output of the active power

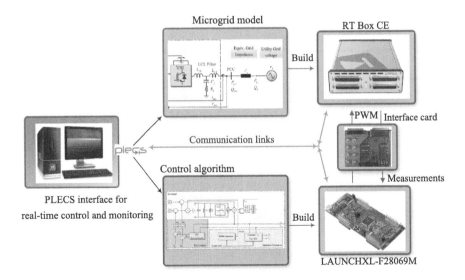

FIGURE 10.12 Block diagram of the hardware-in-the-loop setup.

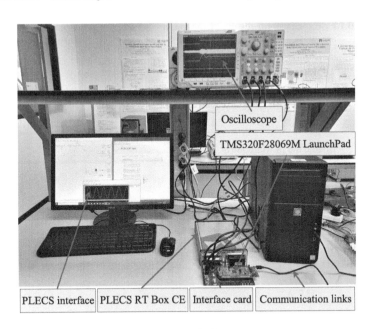

FIGURE 10.13 Photo of the experimental test-bed.

is only $P_{pcc} = 2.52\,\text{kW}$, when the proposed power control is enabled at t_1, the output power is regulated to the desired reference. Similarly, the output active and reactive power are equal to their desired references that are updated to 5, 4 kW, and 1 kVar at t_2, t_3, and t_4, respectively. Figure 10.15 shows the inverter output current corresponding to the desired power changes.

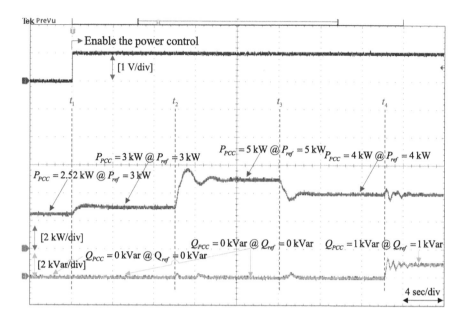

FIGURE 10.14 Experimental results of the active and reactive power of the dVOC-based inverter.

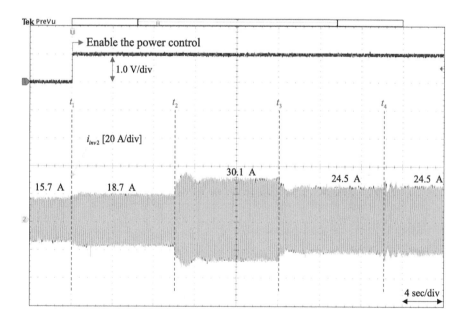

FIGURE 10.15 Experimental results of the output current of the dVOC-based inverter.

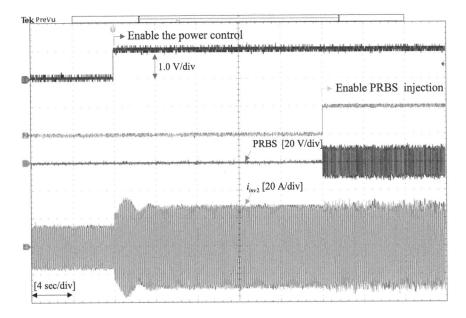

FIGURE 10.16 Experimental results of the inverter output current under PRBS injection for online broadband grid impedance estimation.

10.8.2 ONLINE GIE

In this case study, the PRBS injection by the inverter for online parametric estimation of the grid impedance is investigated. Figure 10.16, the inverter regulates the output power to 5 kW with a unity power factor when the proposed control is enabled at t_1. Injection of the PRBS starts at t_2. As stated previously, the PRBS with an amplitude [−10 V, +10 V] is added to the modulation index of the phase reference voltage. Figure 10.17 shows a zoom-in view of the signals.

Figure 10.18 shows the corresponding measured PCC power including the effects of the PRBS injection on both the active and reactive power.

Figure 10.19 shows the non-parametric estimation of the grid impedance. The selected data for curve fitting are also shown in the same figure.

Figure 10.20 compares the parametric model of the grid impedance with its true value listed in Table 10.2, $Z_{g\ ref} = 0.5 + 0.002\ s$. The obtained parametric model is $Z_{g\ par} = 0.4879 + 0.0002074\ s$. Hence, the estimations errors in the grid resistance and grid inductance are $\Delta R_g\% = -2.43$ and $\Delta L_g\% = 3.69$, respectively. Hence, it is evident that accurate estimation results of the grid impedance are obtained using the methodology for dVOC-based inverters.

10.9   CHAPTER SUMMARY

This chapter proposes the implementation of online GIE for single-phase dispatchable virtual oscillator-controlled GFM inverters operated in grid-connected mode. The online estimation algorithm is embedded into the control loop of the dVOC

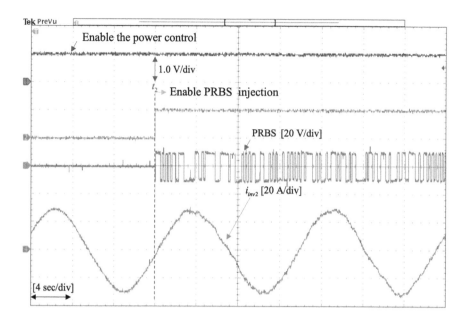

FIGURE 10.17   Zoom-in view of the experimental results presented in Figure 10.16.

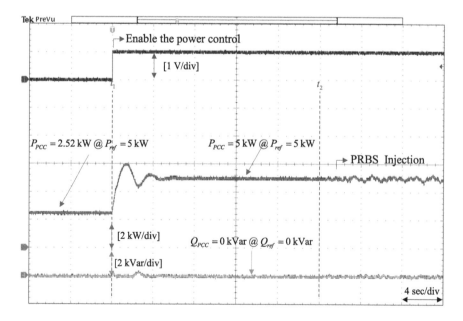

FIGURE 10.18   Experimental results of dispatched power under PRBS injection.

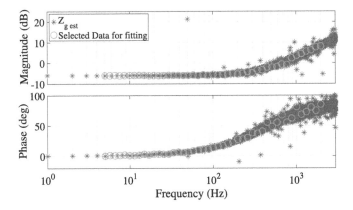

FIGURE 10.19   Experimental results of the non-parametric estimation of grid impedance.

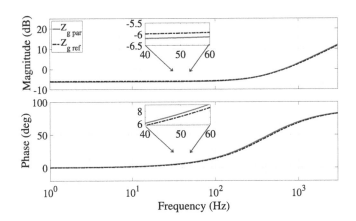

FIGURE 10.20   Experimental results of the parametric estimation of grid impedance and the reference value.

inverter enables online broadband estimation of the equivalent grid impedance seen by the inverter at its PCC based on the PRBS injection. Then, complex curve fitting is employed to obtain the parametric RL model of the equivalent grid impedance. Furthermore, the power control loop has been realized based on the modifying two parameters of the traditional dVOC, the voltage-scaling factor gain k_v, and the current-scaling factor k_i. Through simulations using MATLAB/Simulink and HIL experiments, the dVOC controller is verified to allow injection of the desired active and reactive power with zero power tracking errors. Also, the online GIE algorithm, that is implemented into the control of dVOC-based inverters for the first time, is validated. The obtained high accurate parametric RL model of the grid impedance can be used in further research directions such as online impedance-based stability analysis of dVOC-based inverters.

REFERENCES

[1] Roberto Rosso, Xiongfei Wang, Marco Liserre, Xiaonan Lu, and Soenke Engelken. Grid-forming converters: Control approaches, grid-synchronization, and future trends—A review. *IEEE Open Journal of Industry Applications*, 2: 93–109, 2021.

[2] Josep M Guerrero, Juan C Vasquez, José Matas, Luis García De Vicuña, and Miguel Castilla. Hierarchical control of droop-controlled ac and dc microgrids—A general approach toward standardization. *IEEE Transactions on Industrial Electronics*, 58(1): 158–172, 2010.

[3] Zhan Shi, Jiacheng Li, Hendra I Nurdin, and John E Fletcher. Comparison of virtual oscillator and droop controlled islanded three-phase microgrids. *IEEE Transactions on Energy Conversion*, 34(4): 1769–1780, 2019.

[4] Brian Johnson, Miguel Rodriguez, Mohit Sinha, and Sairaj Dhople. Comparison of virtual oscillator and droop control. In *2017 IEEE 18th Workshop on Control and Modeling for Power Electronics (COMPEL)*, 1–6. IEEE, Stanford, CA, USA, 2017.

[5] Hui Yu, MA Awal, Hao Tu, Iqbal Husain, and Srdjan Lukic. Comparative transient stability assessment of droop and dispatchable virtual oscillator controlled grid-connected inverters. *IEEE Transactions on Power Electronics*, 36(2): 2119–2130, 2020.

[6] Brian B Johnson, Sairaj V Dhople, Abdullah O Hamadeh, and Philip T Krein. Synchronization of parallel single-phase inverters with virtual oscillator control. *IEEE Transactions on Power Electronics*, 29(11): 6124–6138, 2013.

[7] Brian B Johnson, Sairaj V Dhople, James L Cale, Abdullah O Hamadeh, and Philip T Krein. Oscillator-based inverter control for islanded threephase microgrids. *IEEE Journal of Photovoltaics*, 4(1): 387–395, 2013.

[8] Brian B Johnson, Mohit Sinha, Nathan G Ainsworth, Florian D¨orfler, and Sairaj V Dhople. Synthesizing virtual oscillators to control islanded inverters. *IEEE Transactions on Power Electronics*, 31(8): 6002–6015, 2015.

[9] Marcello Colombino, Dominic Groß, Jean-Sébastien Brouillon, and Florian Dörfler. Global phase and magnitude synchronization of coupled oscillators with application to the control of grid-forming power inverters. *IEEE Transactions on Automatic Control*, 64(11): 4496–4511, 2019.

[10] Gab-Su Seo, Marcello Colombino, Irina Subotic, Brian Johnson, Dominic Groß, and Florian Dörfler. Dispatchable virtual oscillator control for decentralized inverter-dominated power systems: Analysis and experiments. In *2019 IEEE Applied Power Electronics Conference and Exposition (APEC)*, 561–566. IEEE, Anaheim, CA, USA, 2019.

[11] David Raisz, Trung Tran Thai, and Antonello Monti. Power control of virtual oscillator controlled inverters in grid-connected mode. *IEEE Transactions on Power Electronics*, 34(6): 5916–5926, 2018.

[12] M. A. Awal and Iqbal Husain. Unified virtual oscillator control for gridforming and grid-following converters. *IEEE Journal of Emerging and Selected Topics in Power Electronics*, 9(4): 4573–4586, 2020.

[13] M. A. Awal, Hui Yu, Hao Tu, Srdjan M. Lukic, and Iqbal Husain. Hierarchical control for virtual oscillator based grid-connected and islanded microgrids. *IEEE Transactions on Power Electronics*, 35(1): 988–1001, 2019.

[14] Puspal Hazra, Ramtin Hadidi, and Elham Makram. Dynamic study of virtual oscillator controlled inverter based distributed energy source. In *2015 North American Power Symposium (NAPS)*, 1–6. IEEE, Charlotte, NC, USA, 2015.

[15] M. A. Awal and Iqbal Husain. Unified virtual oscillator control for gridforming and grid-following converters. *IEEE Journal of Emerging and Selected Topics in Power Electronics*, 9(4): 4573–4586, 2020.

[16] Dominic Groß, Marcello Colombino, Jean-Sébastien Brouillon, and Florian Dörfler. The effect of transmission-line dynamics on grid-forming dispatchable virtual oscillator control. *IEEE Transactions on Control of Network Systems*, 6(3): 1148–1160, 2019.

[17] Tobias Heins, Trung Tran, David Raisz, and Antonello Monti. Power control of andronov-hopf oscillator based distributed generation in gridconnected microgrids. In *International Conference on Engineering Research and Applications*, 675–687. Springer, Thai Nguyen, Vietnam, 2020.

[18] Mihai Ciobotaru, Remus Teodorescu, and Frede Blaabjerg. On-line grid impedance estimation based on harmonic injection for grid-connected PV inverter. In *2007 IEEE International Symposium on Industrial Electronics*, 2437–2442. IEEE, Vigo, Spain, 2007.

[19] Nabil Mohammed and Mihai Ciobotaru. Fast and accurate grid impedance estimation approach for stability analysis of grid-connected inverters. *Electric Power Systems Research*, 207: 107831, 2022.

[20] Mark Sumner, Ben Palethorpe, David W. P. Thomas, Pericle Zanchetta, and Maria Carmela Di Piazza. A technique for power supply harmonic impedance estimation using a controlled voltage disturbance. *IEEE Transactions on Power Electronics*, 17(2): 207–215, 2002.

[21] Karen Tatarian, Micael S Couceiro, Eduardo P Ribeiro, and Diego R Faria. Stepping-stones to transhumanism: An EMG-controlled low-cost prosthetic hand for academia. In *2018 International Conference on Intelligent Systems (IS)*, 807–812. IEEE, Funchal, Portugal, 2018.

[22] Nabil Mohammed, Mihai Ciobotaru, and Graham Town. Online parametric estimation of grid impedance under unbalanced grid conditions. *Energies*, 12(24): 4752, 2019.

[23] Nabil Mohammed, Tamas Kerekes, and Mihai Ciobotaru. An online eventbased grid impedance estimation technique using grid-connected inverters. *IEEE Transactions on Power Electronics*, 36(5): 6106–6117, 2020.

[24] Jian Sun. Impedance-based stability criterion for grid-connected inverters. *IEEE Transactions on Power Electronics*, 26(11): 3075–3078, 2011.

[25] Ke Jia, Tianshu Bi, Bohan Liu, Edward Christopher, David W. P. Thomas, and Mark Sumner. Marine power distribution system fault location using a portable injection unit. *IEEE Transactions on Power Delivery*, 30(2): 818–826, 2014.

[26] Jonas De Kooning, Jan Van de Vyver, Jeroen D. M. De Kooning, Tine L. Vandoorn, and Lieven Vandevelde. Grid voltage control with distributed generation using online grid impedance estimation. *Sustainable Energy, Grids and Networks*, 5: 70–77, 2016.

[27] Henrik Alenius, Roni Luhtala, Tuomas Messo, and Tomi Roinila. Autonomous reactive power support for smart photovoltaic inverter based on real-time grid-impedance measurements of a weak grid. *Electric Power Systems Research*, 182: 106207, 2020.

[28] Mauricio Cespedes and Jian Sun. Adaptive control of grid-connected inverters based on online grid impedance measurements. *IEEE Transactions on sustainable energy,* 5(2): 516–523, 2014.

[29] Rik Pintelon and Johan Schoukens. *System Identification: A Frequency Domain Approach*. Hoboken, NJ: John Wiley & Sons, 2012.

[30] E. C. Levy. Complex-curve fitting. *IRE Transactions on Automatic Control*, AC-4(1): 37–43, 1959.

11 Grid-Forming Inverters Interfacing Battery Energy Storage Systems

Sukumar Kamalasadan, Michael Smith, and Fahim Al Hasnain
The University of North Carolina at Charlotte

CONTENTS

DOI: 10.1201/9781003302520-11

11.1 INTRODUCTION

Increasing the penetration of inverter-based generation in a power system results in reduced system inertia, which can lead to various stability issues. As a result, regulation of voltage and frequency are of considerable concern with the increased usage of non-synchronous generation. Grid-forming (GFM) inverters can provide options to help address these challenges [1,2]. Battery energy storage systems (BESSs) are important for the economic and reliable operation of the grid because of their capability for energy storage, bidirectional energy exchange, and fast output response [3]. With the increasing penetration of renewable energy sources on the grid, where these energy sources usually operate based on maximum power point tracking (MPPT) for economic purposes, the importance of BESSs is becoming more vital. Thus, GFM inverters that are interfaced with BESSs can play a significant role in modern power systems.

To achieve the required system performance, appropriate control strategies are needed for GFM inverters integrated with a BESS. Among several of the control methods for GFM, droop-based control is the most common. The droop-based control method provides regulation of voltage and frequency by mimicking the operation characteristics of a conventional synchronous machine. In this chapter, a detailed architecture of a GFM-based inverter is discussed when a BESS along with a photovoltaic (PV) is connected to the grid. Several load change scenarios, fault scenarios, and grid-connected/islanded operation scenarios are performed for two different test systems. In one test system, the BESS is connected to a single machine infinite bus (SMIB) and another test system contains a larger grid system (IEEE 13 bus) with a BESS integrated. Major BESS components and different BESS chemistry types are also briefly discussed. Finally, a measurement-based identification method, subspace identification, is applied on the DC-Link dynamics to identify modal information for all the cases on both test systems. Topics: inverter architecture (Section 11.2), control architecture (Section 11.3), BESS model (Section 11.4), integrated BESS/GFM inverter model (Section 11.5), large-scale BESS (Section 11.6), application example (Section 11.7), and simulation results (Section 11.8).

11.2 INVERTER ARCHITECTURE

A conventional alternating-current (AC) power system grid mainly consists of synchronous generators, which have various control architectures that are commonly used. As an example, voltage regulation is obtained through excitation control of synchronous generators, whereas frequency control is achieved through governor control. Moreover, the inertia of the prime mover and the rotor of the synchronous generator are required for the stable operation of the power grid, as they help to keep voltage and frequency within acceptable limits during operational events, such as load changes and fault conditions. Synchronous generators also have an automatic voltage regulator mechanism and low output impedance. They also provide short circuit current or fault current contribution capability, which helps to keep the system stable, even after a fault and in clearing the fault [4].

However, with increasing demand for electric power consumption, variable distributed energy resources (DERs) are being integrated with conventional power grids. These DERs are mainly inverter-based resources. Various DERs, such as PV and wind energy production, are becoming more and more common, as a result of reduced costs, availability of resources, and the increasing demand for energy. One major challenge regarding the use of these DERs relates to the intermittent output characteristics of these resources since the output from these DERs is directly dependent on the weather condition in the installation area. As a result of the intermittent nature, DERs often work on the MPPT principle. Consequently, the intermittent nature eventually results in reduced efficiency of the independent source. However, the overall efficiency of DERs can be improved by integration with the utility grid. A microgrid can offer higher reliability, improved power quality, reduced carbon emissions, and a cost-competitive solution over the traditional power distribution system. The connection of the microgrid or DERs to the utility grid is established through an inverter as the interfacing medium [5,6].

Based on the interaction with the grid and controller implementation, inverters can be divided into two main groups: grid-following (GFL) and GFM [1,2,7]. In GFL, inverter-based resources are designed to follow the grid voltage and feed current per the existing voltage. As a result, they are focused on injecting active power into the grid with MPPT, whereas the reactive power supply is minimal. GFL can be further divided into two categories based on reactive power support: grid feeding and grid supporting. In the grid-feeding mode, reactive power support is zero, whereas, in the grid-supporting mode, reactive power is provided to support grid voltage in any deviation. The behavior of the GFL architecture is similar to a current source. On the other hand, in the GFM architecture, inverters work as a voltage source. They are controlled to maintain voltage and frequency. As a result, they can perform in an islanded mode too. Figure 11.1 depicts the different categories of

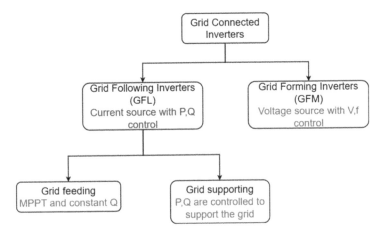

FIGURE 11.1   Classification of grid-connected inverters.

grid-connected inverters, where P is active power, Q is reactive power, V is voltage, and f is frequency.

11.3 CONTROL ARCHITECTURE FOR GFM INVERTER

As mentioned earlier, grid-connected inverters can be classified into two major categories: GFL and GFM. This section details the control architecture for GFL and GFM inverters.

11.3.1 GFL CONTROL

A GFL inverter's behavior can be approximated as a controlled current source that injects current into an existing grid. It has a high impedance in parallel with the controlled current source. Figure 11.2 shows the general control architecture of GFL with the grid connection.

In GFL, the inverter is synchronized with the grid voltage via a phase-locked loop (PLL) mechanism. It measures the voltage at the point of common coupling (PCC) and the phase of the voltage is obtained through the PLL using equation (11.1) [7].

$$\theta_{PLL} = \int \left[\omega_n + (K_p + K_i \int)V_{PCC} \right] \tag{11.1}$$

Here, ω_n is the angular speed (rad/s) and V_{PCC} is the PCC voltage. For the required direct and quadrature axis currents (I_d and I_q), voltage is changed accordingly. Active and reactive power support from GFL is achieved by controlling the injected I_d and I_q current. In GFL, synchronization is necessary throughout the entire time period to maintain synchronization with the grid, as GFL can only deliver real and reactive power since it does not include any voltage and frequency regulation.

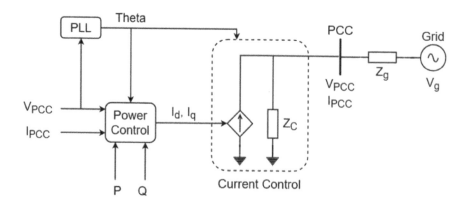

FIGURE 11.2 General grid-following control architecture.

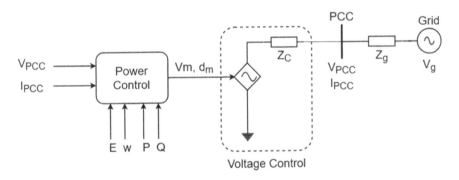

FIGURE 11.3   General grid-forming control architecture.

11.3.2   GFM CONTROL

Microgrids can perform with grid connection, as well, in an islanded mode of operation. While the grid is connection with the microgrid, voltage and frequency are regulated by the synchronous generators of the grid, but in an islanded mode, the microgrid needs to regulate its own voltage and frequency for stable operation. From this perspective, the GFM architecture is introduced and a notable volume of research has been conducted on this relatively new topic. While GFL behavior can be approximated as a controlled current source with a high impedance in parallel, GFM behavior can be approximated as a controlled voltage source with low impedance in series connection with the source. As a result, it can replicate a synchronous machine. For synchronization in the GFM architecture, at the beginning of operation, synchronization can be achieved in a similar fashion to a synchronous machine and further synchronization is not required during normal operation. Figure 11.3 shows the general architecture for GFM with the grid connection.

In GFM, the PCC voltage is measured for regulating the power output. It is able to generate its own voltage and frequency reference in order to perform, even without the presence of the power grid. GFM architectures are recently being used as a solution for weaker grid conditions, as it provides a viable solution by creating its own voltage and frequency reference.

11.4   BESS MODEL

Modern power systems are accommodating an increasing number of variable renewable energy sources with existing grid systems, due to environmental interests, economic issues, and increasing energy consumption [5,8]. Among these sources, solar and wind energy are currently the most common sources used around the world. However, these variable renewable energy sources are strongly dependent upon weather. Hence, their output is not consistent, since weather conditions vary all the time. Another significant concern about renewable energy sources is the reduction in mechanical inertia of the entire power system, which can result in large frequency swings during system disturbances, such as load changes and any type of fault, which in turn can lead the power system to instability [9,10].

As one option to address these problems associated with variable renewable energy sources, hybrid renewable energy systems are considered, which integrate different renewable energy sources in an optimal combination [11–13]. As another viable solution to these problems, various types of energy storage systems are also considered, because of their capability to absorb and release power when needed. Among the different types of energy storage systems, BESSs are being studied extensively [14], because of their energy storage capability, bidirectional energy exchange, geographical independence, and fast output response. In addition, a BESS can also improve power quality, as well as reliability of the system [15,16]. Since the output of a BESS is direct current (DC), an inverter-based solution is required with necessary control (GFM/GFL) for integration with an alternating-current (AC) utility grid.

11.4.1 BESS COMPONENTS

BESS components can be divided into three major parts, namely: components of the battery, components required for grid connection, and components required for reliable system operation.

Battery components consist of a battery pack, which has multiple cells that are arranged in modules to achieve the desired voltage and current capacity, a battery management system (BMS), and a battery thermal management system (B-TMS). The BMS works in a way that maintains proper operation of the BESS within the specified range for the voltage and current, respectively. Also, it protects the BESS in terms of temperature and provides reliable and safe operation of the BESS. The BMS monitors the condition of battery cells, measures their parameters and states, such as state-of-charge (SOC) and state-of-health (SOH), and protects batteries from fires and other hazards. The BMS is also responsible for balancing varying SOC for series connected battery cells. The B-TMS maintains the temperature of the battery cells according to specifications.

Components required for grid connection consist of several items (e.g., power electronics and filters). Power electronics are required for connecting the BESS to the DC bus of the power grid. Filters are required for smoothing the ripples from the output of the BESS. An inverter or a power conversion system (PCS) is required to convert DC power produced by the battery to AC power supplied to grid. For charging or discharging, a bidirectional inverter system is required.

An energy management system (EMS) and thermal management system are required for reliable system operation of BESS. The EMS manages the required power flow and distribution of power flow. It is responsible for energy flow of the battery system. The EMS coordinates between the BMS, PCS, and other components of the BESS, where it is able to efficiently manage power flow after analyzing the data. The thermal management system is responsible for system heating management.

11.4.2 BATTERY CHEMISTRY TYPES

This section provides an overview of the primary battery types for possible consideration with a BESS [3,17,18]. A summary of advantages and disadvantages of different battery chemistry is provided in Table 11.1.

TABLE 11.1
Summary of Typical Battery Chemistry Types Used with a BESS

Chemistry Type	Advantages	Disadvantages
Lead-Acid (PbA) (mainly used in vehicles and other purposes where high values of load current is required)	Simple manufacturing procedure Low cost for manufacturing Low cost per watt hour of usage High discharge current capability Performs irrespective of temperature No cell-wise BMS requirement	Poor weight to energy ratio Slow charging Limited life-cycle Adverse environmental impact
Nickel-Cadmium (Ni-Cd) (mainly used in portable computers, drills, and other small battery-operated devices)	Rechargeable Fast charging with low stress Long shelf life Simple storage and transportation Performs in low temperature Economical pricing	High self-discharge Low cell voltage Environmental impact
Nickel-Metal Hydride (Ni-MH) (combines positives from Ni-Cd batteries with energy storage features of metal alloys)	Higher capacity Less voltage depression Perform at very low temperature Higher energy density	Shorter life span Limited discharge current High self-discharge
Lithium-Ion (Li-Ion) (widely used in all types of electronic devices)	High energy density High load capabilities Maintenance free High shelf life Simple charge algorithm Shorter charge time	Degradation at higher temperatures Requires protection circuit

11.5 INTEGRATED BESS WITH GFM INVERTER MODEL

The section details the model integration of the BESS with the GFM inverter. In this integrated system, a BESS is connected to a grid through DC-Link filters, three-phase switching converters, AC filters, and a transformer. Both GFL and GFM control architectures for the BESS inverter are used in this chapter. A 1 MW solar PV system is also used alongside the BESS, where the PV system is connected to the power grid via a similar fashion to that of the BESS. A one-line diagram of the system used in this chapter is shown in Figure 11.4. The different components of the BESS are discussed in the subsections below.

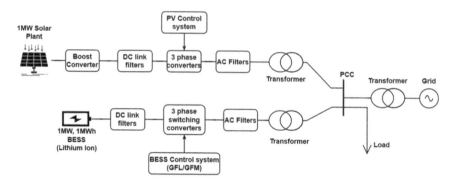

FIGURE 11.4 BESS and PV integration to grid.

11.5.1 INTEGRATED MODEL: BESS

The battery model used in this chapter is a Li-ion model. The no-load voltage of the battery, V_{nl}, is calculated based on the SOC of the battery using a nonlinear equation, which is shown in equation (12.2) [2].

$$V_{nl} = V_0 - V_k \frac{1}{SOC} + A^{-BC(1-SOC)} \tag{11.2}$$

where V_{nl} is the no-load voltage of the battery, V_0 is the battery constant voltage, SOC is the SOC of the battery, V_k is the polarization voltage, C is the capacity of the battery in Ah, and A and B are constants. A, B, and V_k can be tuned for the desired performance from the battery. The BESS has a rated power of 1 MW and a rated capacity of 1 MWh. Each battery cell's voltage is 3.2 V, and each module has four cells in them. There are 72 modules in series. Each cell's current capacity is 14 Ah, and there are 80 parallel connections. The nominal voltage of the BESS is 921.6 V, and the initial SOC is 50%. Aging and temperature effect of the BESS are ignored for this work.

11.5.2 INTEGRATED MODEL: DC/AC FILTER

DC filters are positioned before the converter and AC filters are positioned after the converter. The purpose of both filters is to limit the ripple of the current and to add damping effect. Resistors and inductors of different values are connected in series and capacitors are connected in parallel to accomplish the desired output for both of the filters.

11.5.3 INTEGRATED MODEL: DC-AC CONVERTER

Figure 11.4 shows how the DC-AC converter connects the battery to the grid through filters on both sides. A three-phase two-level switching-type converter is used with the detailed BESS model. Detailed BESS models are better for stability studies and mode identification. A reference voltage signal is passed through pulse width modulation (PWM) to produce a sinusoidal reference signal for the converter. The reference voltage signal is produced by the BESS control system block. The PWM carrier frequency is 2,700 Hz for this model.

11.5.4 Integrated Model: Solar (PV) Panel

A PV panel is also used alongside the BESS for connecting to the grid (Figure 11.4). The PV panel is used to show the BESS performance with a PV panel presented. The PV panel is designed based on a MPPT mode and its rating is 1MW. Each PV cell's open circuit voltage is 51.9 V and the short circuit current is 8.68 A. Each module has 83 cells, and 15 modules are connected in series per string. There are 190 parallel strings in this representative PV model. The PV panel is then connected to a boost converter through DC filters. Then, a three-phase converter is placed for connection between the DC and AC side, accordingly. A PV control block is used, which provides the necessary modulation signal for controlling the PV converter. Then, the PV system is connected to grid through an AC filter and a transformer.

11.6 GRID-CONNECTED/ISLANDED LARGE-SCALE BATTERY ENERGY STORAGE SYSTEMS

A BESS can enhance the power system flexibility and can enable high-level penetration of renewable energy to the grid. A BESS can operate in both grid-connected mode or islanded mode. While connected to grid, a GFL architecture of the inverters is deployed and the BESS follows the grid and supports the grid by supplying or absorbing power, as applicable. While in islanded mode, the GFM architecture of BESS is enabled and the BESS can act as the frequency regulator of the islanded part of the grid. For integration with grid in the transmission network, a large-scale or utility-scale BESS is required. Configuration of a large-scale BESS integration into a transmission network is quite similar to Figure 11.4. Figure 11.5 [19] shows the configuration of a large-scale BESS interconnected at the transmission substation level. Several battery cells are connected in series and parallel to obtain the required

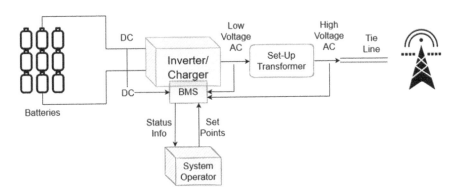

FIGURE 11.5   Large-scale battery energy storage systems.

voltage and current levels and the battery cells are then connected to the inverter, which can operate in both GFL/GFM modes, accordingly. A BMS works alongside the inverter, which can send the status information of several system parameters to the system operator and can also import set points determined by the system operator. The inverter is then connected to the low-voltage side of the AC transformer. The high-voltage side of the AC transformer is connected to transmission network through a tie-line, as depicted in Figure 11.5.

11.7 APPLICATION: HOW TO DEVELOP THE MODEL IN A SOFTWARE PLATFORM (E.G., MATLAB/SIMULINK)

This section provides an example application for the integrated BESS with a GFM inverter, presented in this chapter, which can be performed using a typical software platform (e.g., MATLAB/Simulink). The one-line diagram in Figure 11.4 is modeled in Simulink. A Li-ion BESS model is used alongside a PV panel for integration to the grid. The control algorithm for grid-connected inverters consists of both GFL and GFM architectures. The BESS has a rated power of 1 MW and a rating of 1MWh. The PV panel is rated at 1MW power. The grid generator is rated as 1.3 MVA machine and performs as a swing generator. The terminal voltage of the generator is 20 kV, and the transformer is used as a step-down transformer to transform the voltage to 600 V, which is the PCC voltage. Some fixed load and variable loads are connected to the PCC, which are modeled based on a constant Z model. The PV panel is modeled based on MPPT and the transformer with the PV panel transforms the voltage of the PV panel to 600V and then connects to the PCC. The BESS is modeled in a way that it can perform in both a GFL and GFM mode of control. The nominal voltage of the BESS is 922 V, and the transformer with the BESS transforms the voltage to match the PCC voltage. Voltage and current can be measured at the PCC and other points and buses in the network, as needed. As an example application, Figure 11.6 shows the simulation test case of a BESS integrated with the grid modeled in Simulink.

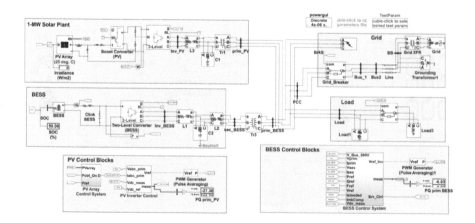

FIGURE 11.6 Simulink model of a BESS integrated with the grid.

11.7.1 Detailed GFL Architecture

The voltage measurement at the PCC and current measurements after the AC filters are passed through an $abc - dq0$ transformation block, which provides dq-axis voltage and current. These dq-axis current and voltage values are then used to calculate the injected active and reactive power per equations (11.3) and (11.4), respectively [2].

$$P = \frac{3}{2}\left(V_d I_d - V_q I_q\right) \tag{11.3}$$

$$Q = \frac{3}{2}\left(V_d I_q - V_q I_d\right) \tag{11.4}$$

A PLL is required to calculate the phase angle θ_{PLL}, which is then transferred to $abc - dq$ blocks and the PWM block. P_{ref} and Q_{ref} are given as inputs to the current reference generator block. This block generates current dq-axis reference set-points I_{dref} and I_{qref} utilizing the calculated powers and reference powers per equations (11.5) and (11.6) [2]. Then, the current closed-loop control block generates dq-axis voltage references $V_{ref_{dq}}$. Both $V_{ref_{dq}}$ and θ_{PLL} signals are then passed through a $dq0 - abc$ transformation block, which provides the abc-reference voltage $V_{ref_{abc}}$. Then, the $V_{ref_{abc}}$ signal is transferred to a PWM scheme to generate the modulation signal for the converter. Figure 11.7 represents the GFL architecture used for this work.

$$I_{dref} = \frac{2}{3}\frac{P_{ref}V_d + Q_{ref}V_q}{V_d^2 + V_q^2} \tag{11.5}$$

$$I_{qref} = \frac{2}{3}\frac{P_{ref}V_q + Q_{ref}V_d}{V_d^2 + V_q^2} \tag{11.6}$$

11.7.2 Detailed GFM Architecture

In the GFM architecture (Figure 11.8), no PLL is required to calculate phase angle. θ_{PLL} is generated using ω_{ref} and then transferred to $abc - dq$ blocks and PWM block.

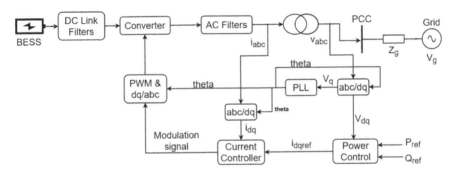

FIGURE 11.7 Grid-following control architecture.

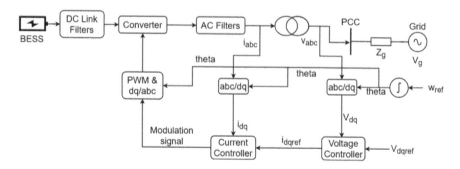

FIGURE 11.8 Grid-forming control architecture.

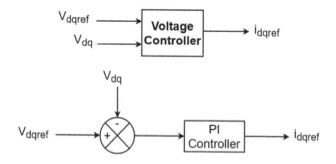

FIGURE 11.9 Voltage controllers.

Voltage and current measurement are collected and passed through $abc - dq0$ blocks to generate dq-axis voltage and current, which are V_{dq} and I_{dq}, respectively. V_{dqref} and V_{dq} signals go into the voltage controller block to generate the I_{dqref} signal for the current controller block. A proportion integral (PI) control block is then used to generate the I_{dqref} signal inside the voltage controller block. Figure 11.9 represents the voltage controller block.

The current control block also has a feedforward block within it. The output from the voltage controller block is passed through the current closed-loop control to obtain the final modulation signal. The feedforward terms should be used to decouple the two axes and should be considered for the difference between the voltages after and before the AC filter. Equations (11.7) and (11.8) [2] represent the feedforward equation, where $V_{dac-filter}$ and $V_{qac-filter}$ are the dq-axes voltage obtained from the voltage measurement at the bus after the AC filter. V_{0d} and V_{0q} are the outputs of the feedforward block. The outputs from feedforward block and the PI controller block are then added to generate the modulation signal. Figure 11.10 represents the current control block with the feedforward terms.

$$V_{0d} = V_{d_{ac-filter}}\left(1 - \omega^2 L_f C_f\right) + I_{dref}R_f - I_{qref}\omega L_f - V_{q_{ac-filter}}\omega R_f C_f \qquad (11.7)$$

$$V_{0q} = V_{q_{ac-filter}}\left(1 - \omega^2 L_f C_f\right) + I_{dref}R_f + I_{dref}\omega L_f + V_{d_{ac-filter}}\omega R_f C_f \qquad (11.8)$$

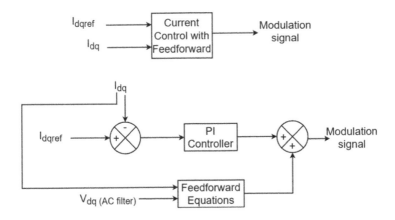

FIGURE 11.10   Current controllers.

The outputs of the current controller are the modulation signals, which are transformed back to the abc-reference frame to obtain the sinusoidal control signals for
the PWM scheme of the converter.

A droop controller is used to generate the ω_{ref} and V_{dqref} signals, which are used
with the voltage control blocks and phase angle calculation. Droop control is also
used for maintaining voltage and frequency within acceptable limits. Figure 11.11
represents the droop control block, which is used for this work. An active power
controller is used for generating ω_{ref} and a reactive power controller is used for generating V_{dqref}.

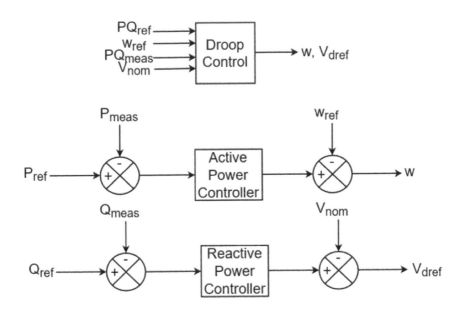

FIGURE 11.11   Droop control.

11.8   SIMULATION RESULTS

Two test systems are used for this chapter. In Test System 1, the BESS and PV systems are connected to a generator that works as a swing generator for this system (see Figure 11.4). In Test System 2, the BESS and PV systems are connected to a larger power system, which is an IEEE 13 bus system (see Figure 11.21). Both test systems are simulated in MATLAB/Simulink. The discrete timestep is 40 μs. For each test system, four different cases or events are studied. For all cases, the DC-Link voltage dynamics measurement is captured for stability studies.

11.8.1   TEST SYSTEM 1: BESS WITH A SINGLE MACHINE

For this test system (Test System 1), a BESS along with a PV system is connected to a single grid machine, which has a rating of 1.3 MVA (Figure 11.4). The phase-to-phase RMS voltage is 20 kV, which is then passed through a step-down transformer and transformed to 600 V. The BESS and PV are connected to the PCC via transformers, where the PCC voltage is 600 V. The total load connected at the PCC is 600 kW and 100 kVAR, respectively. Some variable load is also connected to create a load change event or step change in load during the study.

11.8.2   CASES ON TEST SYSTEM 1

11.8.2.1   Case 1: GFL and GFM

Case 1 (Test System 1) shows the transformation of the grid-connected inverter from GFL mode to GFM mode. For this test case, the total simulation period is 6 s. The total load connected at PCC is 600 kW and 100 kVAR, respectively. The grid-connected generator is supplying the load initially (from 0 to 1 s). No PV or BESS supply is present initially. From 1 to 3 s, GFL mode is activated with a reference power of −200 kW at the BESS, where the BESS is charging by 200 kW. Grid-supplied power is increased by 200 kW to charge the BESS. From 3 to 6 s, GFM mode is activated. The grid is islanded and the BESS starts to supply the demanded load. The PV irradiance is 0 W/m^2 for the total simulation period and the output power from the PV is also 0 kW throughout the entire simulation period.

Figure 11.12 shows the power shared by all sources at the PCC. Figure 11.13 depicts the DC-Link voltage dynamics from the BESS for this case.

The PCC voltage and frequency are shown in Figure 11.14. From this figure, in GFL mode, the PCC voltage remains at 600 V and the PCC frequency remains at 60 Hz, but in GFM mode, the frequency varies between 59.80 and 59.85 Hz.

11.8.2.2   Case 2: Step Change in Load in GFM with PV Supply

Case 2 (Test System 1) considers a step change in the load, while in GFM mode with the PV supply. For this test case, the total simulation period is 6 s. The total initial load is 600 kW and 100 kVAR at the beginning of simulation. Irradiance of the PV plant is 200 W/m^2 and the PV system is supplying 200 kW throughout the entire simulation. Initially, the grid is supplying the additional 400 kW and 100 kVAR (from 0 to 1 s). From 1 to 6 s, GFM mode is activated. The BESS is supplying the 400 kW

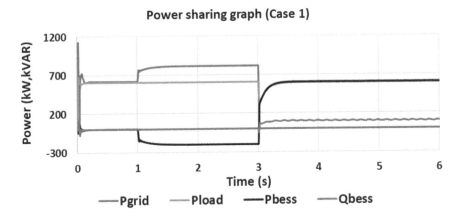

FIGURE 11.12   Power-sharing graph at PCC for Case 1 (Test System 1).

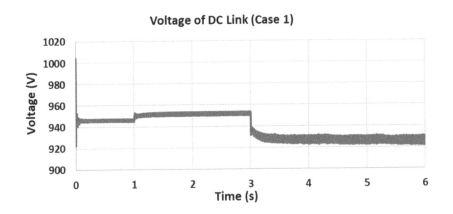

FIGURE 11.13   DC-Link voltage dynamics for Case 1 (Test System 1).

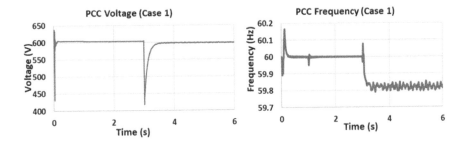

FIGURE 11.14   PCC voltage and frequency for Case 1 (Test System 1).

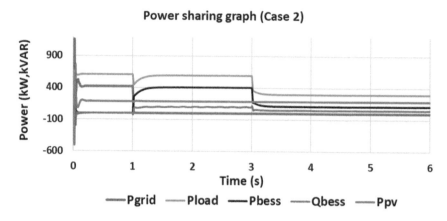

FIGURE 11.15   Power-sharing graph at PCC for Case 2 (Test System 1).

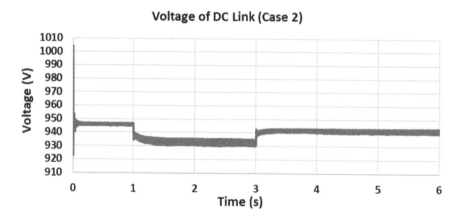

FIGURE 11.16   DC-Link voltage dynamics for Case 2 (Test System 1).

and 100 kVAR load, instead of the grid generator. At 3 s, half of the active and reactive load is disconnected. As a result, the BESS is supplying 100 KW and 50 kVAR.

Figure 11.15 shows the power shared by all sources at the PCC. Figure 11.16 represents DC-Link voltage dynamics from the BESS.

The PCC voltage and frequency are shown in Figure 11.17. From this figure, initially, the PCC voltage remains at 600 V, and the PCC frequency remains at 60 Hz. From 1 to 3 s, the PCC frequency is around 59.875 Hz, and after 3 s, the PCC frequency is around 59.97 Hz.

11.8.2.3   Case 3: Fault at PCC in GFM with PV Supply

Case 3 (Test System 1) considers a fault at the PCC, while in GFM mode with the PV supply. For this test case, the total simulation period is 6 s. The total load is 600 kW and 100 kVAR at the beginning of the simulation. The PV system is supplying 200 kW

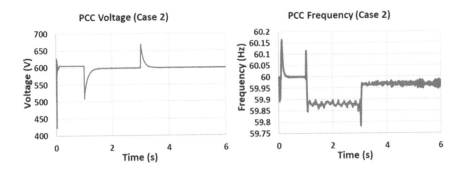

FIGURE 11.17 PCC voltage and frequency for Case 2 (Test System 1).

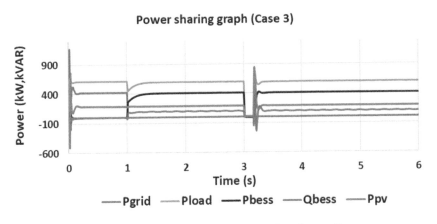

FIGURE 11.18 Power-sharing graph at PCC for Case 3 (Test System 1).

throughout the entire simulation. The power grid is initially supplying 400 kW and 100 kVAR. From 1 to 6 s, GFM mode is activated. The BESS supplies 400 kW and 100 kVAR to supply the load. At 3 s, a three-phase fault is applied for 0.167 s. At 3.167 s, the fault is cleared and the whole system goes back to previous state. Three-phase fault is applied to check stability under most stress.

Figure 11.18 shows the power shared by all sources at the PCC. Figure 11.19 depicts DC-Link voltage dynamics from the BESS.

The PCC voltage and frequency are shown in Figure 11.20. From this figure, initially, the PCC voltage remains at 600 V and the PCC frequency remains at 60 Hz. From 1 to 6 s, the PCC frequency is around 59.87 Hz.

For all cases considered with Test System 1, the DC-Link dynamics are measured and a mode identification method, called subspace identification [20], is applied on the measurement data to calculate the mode and damping ratio. The modal and damping ratio information helps to identify any unstable modes and apply proper control mechanism to stabilize the system [21–23]. The summary of identified modes of all cases is summarized in Table 11.2.

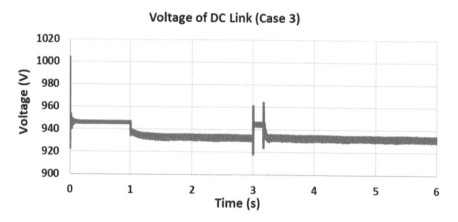

FIGURE 11.19 DC-Link voltage dynamics for Case 3 (Test System 1).

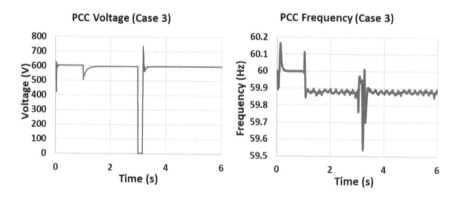

FIGURE 11.20 PCC voltage and frequency for Case 3 (Test System 1).

TABLE 11.2

Summary of Electromechanical Mode Identification for Test System 1

Cases	Steady-State Mode (Hz) and Damping Ratio (%)	GFL/GFM Mode (Hz) and Damping Ratio (%)	Load Change/Fault Mode (Hz) and Damping Ratio (%)
Case 1	0.086 and 12.098	0.155 and −0.0008	0.155 and −0.0007
Case 2	0.086 and 12.098	0.155 and −0.0012	0.155 and −0.0076
Case 3	0.086 and 12.098	0.155 and −0.0012	0.155 and −0.0164

11.8.3 TEST SYSTEM 2: BESS AND PV INTEGRATED WITH IEEE 13 BUS

For Test System 2, the BESS and PV systems are connected to bus 634 of the IEEE 13 bus system (see Figure 11.21). The phase-to-phase RMS voltage is 480 V. The BESS and PV systems are connected to the PCC via transformers, where the PCC

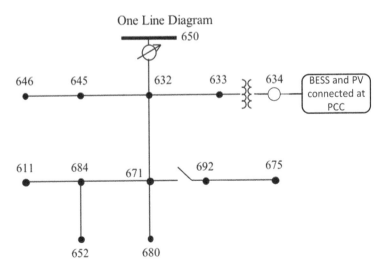

One Line Diagram

FIGURE 11.21   IEEE 13 bus system.

voltage is 480 V. The total load connected to the PCC is 400 kW and 290 kVAR, respectively.

11.8.4   CASES ON TEST SYSTEM 2

11.8.4.1   Case 1: GFL and GFM

Case 1 (Test System 2) considers the transformation of the grid-connected inverter with the IEEE 13 bus system from GFL mode to GFM mode. The total simulation period is 5 s. The total load is 400 kW and 290 kVAR, respectively. Initially, from 0 to 1 s, the grid is supplying the load. From 1 to 3 s, GFL mode is activated with a reference power of −200 kW at the BESS, where the BESS is charging by 200 kW. The grid-supplied power is increased by 200 kW to charge the BESS. From 3 to 5 s, GFM mode is activated. The grid is islanded and the BESS starts to supply the load. The PV system irradiance is 0 W/m², and the output of the PV plant is 0 kW for the entire simulation period.

Figure 11.22 shows the power shared by all sources at the PCC. Figure 11.23 depicts the DC-Link voltage dynamics from the BESS. In GFL mode, the PCC voltage remains at 480 V, and the PCC frequency is 60 Hz, whereas the PCC frequency is around 59.95 Hz in GFM mode.

11.8.4.2   Case 2: Step Change in Load in GFM with PV Supply

Case 2 (Test System 2) considers a step change in the load, while in GFM mode with the PV supply and IEEE 13 bus system. In this test case, the total simulation period is 5 s. The total load is 400 kW and 290 kVAR at the beginning of the simulation. The PV system is supplying 50 kW throughout the entire simulation. The power grid is supplying 350 kW and 290 kVAR at the start. From 1 to 5 s, GFM mode is activated. The BESS starts supplying the 350 kW and 290 kVAR load. At 3 s, half of the load

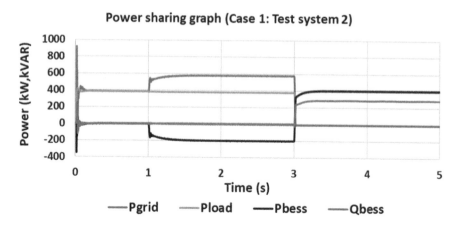

FIGURE 11.22   Power-sharing graph at PCC for Case 1 (Test System 2).

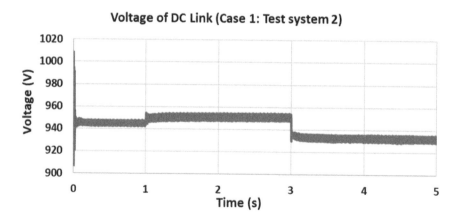

FIGURE 11.23   DC-Link voltage dynamics for Case 1 (Test System 2).

is disconnected. After 3 s, the BESS is supplying the load, which is 150 kW and 145 kVAR.

Figure 11.24 shows the power shared by all sources at the PCC. Figure 11.25 depicts the DC-Link voltage dynamics from the BESS. Initially, the PCC voltage remains at 480 V, and the PCC frequency remains at 60 Hz. From 1 to 3 s, the PCC frequency is around 59.95 Hz, and after 3 s, the PCC frequency is around 59.97 Hz.

11.8.4.3   Case 3: Fault at PCC in GFM with PV Supply

Case 2 (Test System 2) considers a fault at the PCC, while in GFM mode with the PV supply and IEEE 13 bus system. In this case, the total simulation period is 5 s. The total load is 400 kW and 290 kVAR at the beginning of the simulation. The PV is supplying 200kW throughout the entire simulation. Initially, the power grid is supplying

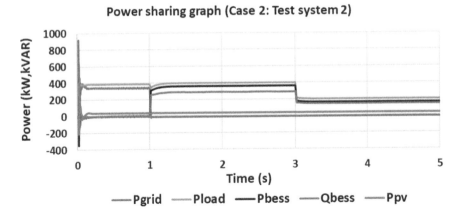

FIGURE 11.24   Power-sharing graph at PCC for Case 2 (Test System 2).

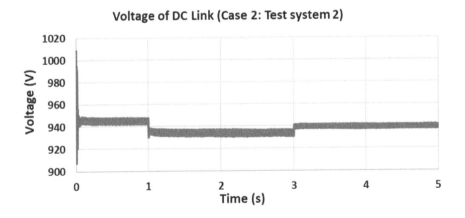

FIGURE 11.25   DC-Link voltage dynamics for Case 2 (Test System 2).

200 kW and 290 kVAR. From 1 to 5 s, GFM mode is activated. The BESS supplies 200 kW and 290 kVAR, respectively, to supply the load. At 3 s, a three-phase fault is applied for 0.167 s at the inverter side. At 3.167 s, the fault is cleared and the power system goes back to the previous state.

Figure 11.26 shows the power shared by all sources at the PCC. Figure 11.27 depicts DC-Link voltage dynamics from the BESS. Initially, the PCC voltage remains at 480 V, and the PCC frequency remains at 60 Hz. From 1 to 5 s, the PCC frequency is around 59.975 Hz.

For all of the test cases considered with Test System 2, the DC-Link dynamics are measured. The associated modes and damping ratios are calculated through a subspace identification method. The summary of the identified modes for all cases of Test System 2 is summarized in Table 11.3.

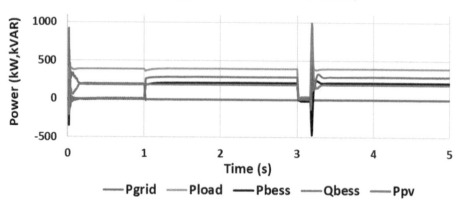

FIGURE 11.26   Power-sharing graph at PCC for Case 3 (Test System 2).

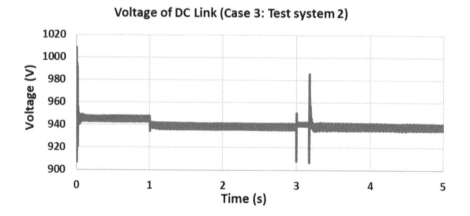

FIGURE 11.27   DC-Link voltage dynamics for Case 3 (Test System 2).

TABLE 11.3

Summary of Electromechanical Mode Identification for Test System 2

Cases	Steady-State Mode (Hz) and Damping Ratio (%)	GFL/GFM Mode (Hz) and Damping Ratio (%)	Load Change/Fault Mode (Hz) and Damping Ratio (%)
Case 1	0.081 and 40.09	0.155 and 0.1153	0.155 and 0.0105
Case 2	0.081 and 40.09	0.155 and 0.0175	0.155 and −0.0035
Case 3	0.081 and 40.09	0.155 and 0.0381	0.155 and 0.01

11.9   CHAPTER SUMMARY AND CONCLUSION

This chapter presents a BESS, along with a PV system, that is connected to a power grid through both GFL and GFM control architectures. This chapter also presents the basic fundamental differences between GFL and GFM architectures for grid-connected inverters. GFM architectures for controlling grid-connected inverters are an emerging technology and can be utilized for maintaining stability in the grid by providing voltage and frequency support, especially in the weaker parts of the grids. With the increasing penetration of variable distributed energy resources into present AC utility grids, more inverter-based technologies are being used, where GFM with a BESS can offer a viable solution to enable integration of these inverter-based resources with the existing power grid by addressing performance challenges (e.g., reliability and stability). In this chapter, integration of a BESS is shown with a single machine, as well as a larger power grid system (IEEE 13 bus system), to showcase the capability of using GFM in larger grids. Also, oscillatory modes can be identified from measurement data collected from SCADA or phasor measurement units (PMUs) to find unstable modes and damping ratios, which can help to build a stability control architecture.

REFERENCES

[1] Dayan B. Rathnayake, Milad Akrami, Chitaranjan Phurailatpam, Si Phu Me, Sajjad Hadavi, Gamini Jayasinghe, Sasan Zabihi, and Behrooz Bahrani. Grid forming inverter modeling, control, and applications. *IEEE Access*, 9: 114781–114807, 2021.

[2] Mostafa Farrokhabadi, Sebastian König, Claudio A. Cañizares, Kankar Bhattacharya, and Thomas Leibfried. Battery energy storage system models for microgrid stability analysis and dynamic simulation. *IEEE Transactions on Power Systems*, 33(2): 2301–2312, 2018.

[3] K.C. Divya and Jacob Østergaard. Battery energy storage technology for power systems—an overview. *Electric Power Systems Research*, 79(4): 511–520, 2009.

[4] Prabha Kundur, Neal J Balu, and Mark G Lauby. *Power System Stability and Control*, volume 7. McGraw-hill, New York, 1994.

[5] Lubna Mariam, Malabika Basu, and Michael F Conlon. A review of existing microgrid architectures. *Journal of Engineering*, 2013, 2013, Article ID 937614, 8 pages.

[6] Lubna Mariam, Malabika Basu, and Michael F Conlon. Microgrid: Architecture, policy and future trends. *Renewable and Sustainable Energy Reviews*, 64: 477–489, 2016.

[7] Xikun Fu, Jianjun Sun, Meng Huang, Zhen Tian, Han Yan, Herbert Ho-Ching Iu, Pan Hu, and Xiaoming Zha. Large-signal stability of grid-forming and grid-following controls in voltage source converter: A comparative study. *IEEE Transactions on Power Electronics*, 36(7): 7832–7840, 2021.

[8] Olalekan Ogundairo, Sukumar Kamalasadan, Anuprabha R. Nair, and Michael Smith. Oscillation damping of integrated transmission and distribution power grid with renewables based on novel measurement-based optimal controller. *IEEE Transactions on Industry Applications*, 58(3): 4181–4191, 2022.

[9] Pragya Nema, R.K. Nema, and Saroj Rangnekar. A current and future state of art development of hybrid energy system using wind and pv-solar: A review. *Renewable and Sustainable Energy Reviews*, 13(8): 2096–2103, 2009.

[10] Sunanda Sinha and S.S. Chandel. Review of recent trends in optimization techniques for solar photovoltaic–wind based hybrid energy systems. *Renewable and Sustainable Energy Reviews*, 50: 755–769, 2015.

[11] Ricardo Luna-Rubio, Mario Trejo-Perea, D. Vargas-Vázquez, and G.J. Ríos-Moreno. Optimal sizing of renewable hybrids energy systems: A review of methodologies. *Solar Energy*, 86(4): 1077–1088, 2012.

[12] Subho Upadhyay and M.P. Sharma. A review on configurations, control and sizing methodologies of hybrid energy systems. *Renewable and Sustainable Energy Reviews*, 38: 47–63, 2014.

[13] Prem Prakash and Dheeraj K. Khatod. Optimal sizing and siting techniques for distributed generation in distribution systems: A review. *Renewable and Sustainable Energy Reviews*, 57: 111–130, 2016.

[14] Yuqing Yang, Stephen Bremner, Chris Menictas, and Merlinde Kay. Battery energy storage system size determination in renewable energy systems: A review. *Renewable and Sustainable Energy Reviews*, 91: 109–125, 2018.

[15] Nirmal-Kumar C Nair and Niraj Garimella. Battery energy storage systems: Assessment for small-scale renewable energy integration. *Energy and Buildings*, 42(11): 2124–2130, 2010.

[16] Andreas Poullikkas. A comparative overview of large-scale battery systems for electricity storage. *Renewable and Sustainable Energy Reviews*, 27: 778–788, 2013.

[17] Jaephil Cho, Sookyung Jeong, and Youngsik Kim. Commercial and research battery technologies for electrical energy storage applications. *Progress in Energy and Combustion Science*, 48: 84–101, 2015.

[18] Xing Luo, Jihong Wang, Mark Dooner, and Jonathan Clarke. Overview of current development in electrical energy storage technologies and the application potential in power system operation. *Applied Energy*, 137: 511–536, 2015.

[19] Thomas Bowen, Ilya Chernyakhovskiy, and Paul L. Denholm. Grid-scale battery storage: frequently asked questions. Technical report, National Renewable Energy Lab. (NREL), Golden, CO (United States), 2019.

[20] Peter Van Overschee and B.L. De Moor. *Subspace Identification for Linear Systems: Theory—Implementation—Applications*. Springer Science & Business Media, New York, 2012.

[21] Fahim Al Hasnain, Sheikh Jakir Hossain, and Sukumar Kamalasadan. Investigation and design of a measurement based electro-mechanical oscillation mode identification and detection in power grid. In *2021 IEEE Power & Energy Society General Meeting (PESGM)*, 1–5, IEEE, Washington, DC, USA, 2021.

[22] Fahim Al Hasnain, Amirreza Sahami, and Sukumar Kamalasadan. An online wide-area direct coordinated control architecture for power grid transient stability enhancement based on subspace identification. *IEEE Transactions on Industry Applications*, 57(3): 2896–2907, 2021.

[23] Fahim Al Hasnain, Sheikh Jakir Hossain, and Sukumar Kamalasadan. A novel hybrid deterministic-stochastic recursive subspace identification for electromechanical mode estimation, classification, and control. *IEEE Transactions on Industry Applications*, 57(5): 5476–5487, 2021.

12 Operation of Grid-Forming Inverters in Islanded Mode

Habib Ur Rahman Habib

University of Engineering and Technology Taxila

Durham University

CONTENTS

DOI: 10.1201/9781003302520-12

253

12.1 INTRODUCTION

Energy storage systems (ESSs) are required by microgrids (MGs) based on renewable energy resources (RERs) to tackle the power unbalancing that is primarily brought on by the variable and intermittent behaviour of RERs. Battery applications are nevertheless constrained by greater costs and a life period that is dependent on charging and discharging cycles. By managing the charging and discharging based on state of charge (SOC), an optimum operation control can lengthen the life of a battery bank. It is inevitable that the control for achieving other control goals must be coordinated with the optimal operating control of the battery bank in an MG based on RERs. For instance, creating the control scheme for the MG will inevitably include thinking about how to increase the MG's efficiency under various system situations. Consequently, the RERs-based MG optimum control issue has many objectives.

One of the main objectives of future intelligent or smart MGs is to switch to 100% RER-based electricity generation. The integration of RERs into the MG is made more difficult by the diverse, intermittent, and frequently extensively geographically spread nature of RERs. The largest issue holding up the transition to the integration of 100% pure green renewable energy into the MG system is the scaled-back RER generation systems due to their sporadic nature. Despite this, it is still possible to govern a 100% pure green MG system with a sufficient capacity ESS to run effectively by creating a reliable control system for stabilising the erratic and intermittent RERs.

Voltage unbalance is one of the primary power quality problems in three-phase MG systems in low-voltage distribution networks [1]. It has a crucial impact on voltage-sensitive loads at the distribution level, which is mostly brought on by unbalanced and non-linear loads [2]. The concept of unbalance includes both the phase angle variations between the phases as well as the amplitude variances between the three phases. One of the main factors contributing to an unbalanced voltage is the uneven distribution of single-phase loads. In the past, distribution lines and an active filter were used to reduce the severity of voltage imbalances [3–5]. Another option is to inject negative sequence current to balance out unbalanced [1,6,7] and regulate distribution line flow. However, each of these methods calls for additional compensation hardware, which could raise MG's overall investment costs. It is economically feasible to use distributed generation (DG) units to reduce the voltage unbalance of MG in the MGs framework due to the high interfacing converter usage, as demonstrated in the work in [8–12].

For MGs to operate dependably and economically, power quality is crucial. Additionally, total harmonic distortion (THD) is a widely used benchmark statistic for evaluating the quality of power. Currently, RERs are more frequently incorporated into MG designs and coupled via power converters. By adding undesirable harmonics to load current and bus voltage, the operation of switching converters and non-linear loads changes the power characteristics of distribution networks. Additionally, for MGs connected to major grids, undesirable harmonics from them may significantly degrade power quality at the point of common coupling (PCC). Effective control methods are necessary to combat the harmonic issue MGs face in order to lower the overall THD to less than 5% (per the IEEE 519 standard), as conventional harmonic control techniques are very expensive and ineffective for MGs.

By using optimal control strategies in converters, it is possible to reduce the THD of MGs to the lowest level achievable without spending a lot of money compared to the current conventional systems. Model predictive control (MPC), for instance, offers the benefits of multivariable control with quick and reliable dynamic response. The demanding requirements of voltage balancing and low harmonics may be managed simultaneously if MPC is employed in the converters of MGs. In addition, MPC is easier to implement than traditional control approaches.

For automatically improving MG operation, it is advisable to create a low-cost integrated energy management system (EMS). The main challenge in developing a strong and stable integrated EMS of MG operation is to select a specific region with appropriate architecture and rules in light of the increase in cyberattacks. On the other hand, as an increasingly viable alternative to large, conventional central-ised power plants for the electrification of rural areas, the integration of RER-based DGs in the main grid or in grid-forming islanded MG is growing. As a result of the high RER penetration and the switching of MG operation modes, the steady control operation of MGs is becoming more difficult.

The coordinated control of DGs and loads for stable operation, device protection, load scheduling, energy management (ESS efficiency and longevity, usage of renew-able resources, etc.), and enhancement of power quality should all be included in the integrated EMS of MGs.

The integrated EMS with coordinated control methods suggested in this chapter makes the control structure simpler and the parameter design easier. In the proposed integrated EMS, the power switches are directly controlled by the complete state-variable MPC in order to achieve quick tracking of the reference signals. The archi-tecture of the suggested system is appropriate for achieving the best performance of numerous control objectives, the capability of simple DG expansion and upgrade in plug-and-play mode, the robustness of operation, and the resilience to communica-tion failure.

The simulation model for the MG with solely RER-based DGs has been created in this chapter using the MATLAB/Simulink® environment. The coordinated operation control of the RERs in the MG has since been created for the integrated EMS. Finite control set model predictive control (FCS-MPC)-based coordinated control has been suggested in the integrated EMS for the DGs connected by interlinking power converter (IPC). The MPC-based IPC is designed to enhance system performance in terms of power quality, such as voltage regulation and THD management under steady-state and transient conditions of PV-wind generation and sudden load shifts.

The analysed MG's schematic diagram with the suggested integrated EMS is shown in Figure 12.1. The suggested integrated EMS incorporates each coordinated control unit. As a voltage source inverter (VSI) and active front-end rectifier, the IPC is regulated in both directions. Through the integrated EMS, the switching block is used to control the MG system in both stand-alone and grid-connected modes.

It can be seen that the single inverter current references are only provided by an outer power controller without a droop controller. Due to single inverter without parallel operation, the system will operate and the inverter will form a grid in the islanded mode. The following sections of this chapter go into great detail about the specific functions of each block.

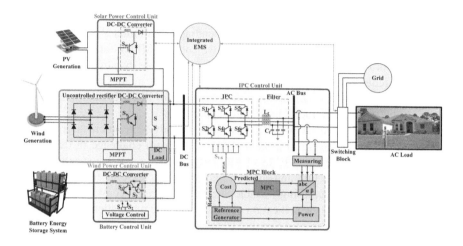

FIGURE 12.1 Diagrammatic representation of the investigated MG and the suggested Integrated EMS.

12.2 CONTROL MODELS OF THE STUDIED MG COMPONENTS

The development of the suggested integrated EMS of the examined MG is fundamentally dependent upon the control models of the photovoltaic (PV), wind turbine (WT), and ESS. The ensuing subsections provide illustrations of these models.

12.2.1 MODELLING OF WTs WITH CONTROL STRUCTURE

The control structure and block diagram of the proposed MG's WT, which is based on a permanent magnet synchronous generator (PMSG), are shown in Figure 12.2. The WT's mathematical modelling is provided in chapter. Where $A = \pi R^2$ is the representation of turbine swept area in m²; WT characteristics are $C_1 = 0.220$, $C_2 = 116$, $C_3 = 0.400$, $C_4 = 5.000$, $C_5 = 12.500$, $C_6 = 0.000$. By Substitution, we get [13]:

$$P_{MPPT}^{WG} = \frac{\rho \pi R^5 k_g^3 C_{p\max}}{2\lambda^3} . \omega_D^3 \qquad (12.1)$$

where k_g is the high-speed shaft's stiffness coefficient [14]. Pitch angle control and the maximum power point tracking (MPPT) approach are both simulated in order to regulate the active power of WT. Using the MPPT model, the dynamic reference power is calculated from the current rotor speed (ω_D). The following equation describes how to automatically attain the ideal rotor speed [13]:

$$2H_D . \omega_D . \frac{d\omega_D}{dt} = P_{mec} - P_{MPPT}^{WG} \qquad (12.2)$$

where P_{MPPT} is the load power supplied by the wind generator at MPPT, P_{mec} is the mechanical power, and H_D is the turbine's inertia constant [15]. The mathematical

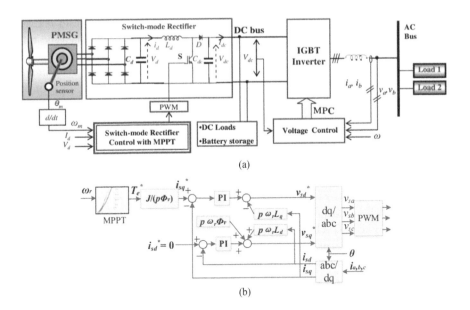

(a)

(b)

FIGURE 12.2   (a) Control scheme for WT based on PMSG. (b) System block schematic for the WT controller.

representation of PMSG WT in the rotating reference frame d-q is built upon the equations listed below [16]:

$$\begin{cases} V_d = -R_S I_d - L_d \dfrac{dI_d}{dt} + \omega_e \varphi_q \\[3mm] V_q = -R_S I_q - L_q \dfrac{dI_q}{dt} + \omega_e \varphi_d \end{cases} \tag{12.3}$$

$$\omega_e = p.\omega_m \tag{12.4}$$

where the q-axis and d-axis currents and stator inductances, respectively, are Iq, Id and Lq, Ld; stator resistance, or R_S; p is the number of pole pairs, and e, m are the electrical and mechanical angular speeds, respectively. In the coupling electromagnetic form, q and d are flux connections in the q- and d-axes, respectively, and are stated as follows [16]:

$$\begin{cases} \varphi_d = L_d I_d + \varphi_f \\[2mm] \varphi_q = L_q I_q \end{cases} \tag{12.5}$$

where φ_f is the flux linkage of a permanent magnet. The following is an expression for the electromagnetic torque [16]:

$$T_{em} = \frac{3}{2} p \left[\left(L_d - L_q \right) I_d I_q + I_q \varphi_f \right] \tag{12.6}$$

Hence, it is important to maintain the shaft speed at an optimal value to deliver maximum P_{mec} under a specific wind speed. The PMSG is activated when the turbine turns in order to create AC voltage in the stator windings. It is noticed that the optimum generator T_e-ω_r characteristic may be acquired dependent on the optimum WT P_m-ω_m curve. The DC bus is then connected to the PMSG stator through an electrical interface. A diode rectifier with a chopper is typically used for low-rated WTs. Voltage source pulse width modulation (PWM) converters are often utilised for larger power ratings. A control approach without a speed sensor is used to maximise wind power extraction. The algorithmic control system of PMSG-based WT is shown in Figure 12.2b. The torque reference is obtained from the optimum Te-ωr characteristic as per the rotor speed. The q part of the stator current is determined from inertia, the number of pole pairs, and flux linkage. From that point on, the PI controller will be used to calculate the necessary converter voltage using PMSG dynamic equations [17]. The difference between the generator and turbine torques is used to compute the generator's acceleration or deceleration. The generator will run in acceleration mode when the generator speed is below the ideal level and the turbine torque is greater than the generator torque. If the generator speed is higher than the recommended speed, on the other hand, the generator will operate in deceleration mode. As a result, both torques reach their ideal values at any given wind speed, and the WT operates at its peak power (MPP). The power balance and battery SOC can determine whether the wind generator is operated on-MPPT or off-MPPT.

12.2.2 MODELLING A PV SYSTEM WITH A CONTROL STRUCTURE

The DC-DC converter connects the PV array to the DC bus. The solar-powered PV modules are connected in series to form PV strings, which are then connected in parallel to form a PV array. The output current/voltage characteristics of the PV mathematical modelling were already largely derived in [18]. Under different irradiance/temperature conditions, it is frequently investigated how the maximum power points (MPPs) of the P-V curves correspond to various voltages. Similar to this, the PV panel output fluctuates with different terminal voltages under specific irradiance. Consequently, the PV voltage needs to be properly balanced in order to implement MPP tracking. The converter can be managed to achieve this. As was previously said, the PV voltage needs to be balanced by adjusting the boost converter in order to produce the most power possible under various irradiance conditions. This effort makes use of an incremental conductance (IC)-based MPPT technique [19] with improved steady-state and dynamic performance under rapidly changing atmospheric conditions. Be aware that the MG power balance and battery SOC can determine whether the converter is controlled on-MPPT or off-MPPT. Figure 12.3a's PV cell equivalent circuit is represented as the following [20]:

$$I_{ph} = I_D + I_{pv} + \frac{V_{pv} + R_{se} + I_{pv}}{R_{sh}} \tag{12.7}$$

$$I_D = I_o \left(e^{\frac{V_D}{V_{Temp}}} - 1 \right) \tag{12.8}$$

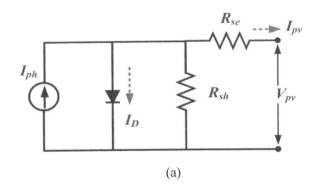

(a)

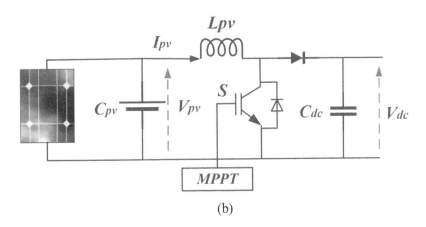

(b)

FIGURE 12.3 (a) The PV cell's equivalent circuit. (b) An arrangement of a solar PV system.

$$V_D = V_{pv} + R_{se} \times I_{pv} \qquad (12.9)$$

$$V_{Temp} = \frac{N \times K \times T_m}{Q} \qquad (12.10)$$

where I_{ph} is the photon or light-generated current of PV cell in Ampere, I_D is the diode current, I_{pv} is the PV cell current at nominal conditions, V_{pv} is the PV cell voltage, R_{se} is the series resistance, R_{sh} is the shunt resistance, I_o is the diode saturation current, V_{temp} is the temperature voltage, V_D is the voltage across the diode, N is the quality factor of material at junction, K is Boltzman constant ($1.3806503 \times 10^{-19}$ J/K) [21], T_m is the measured temperature in Kelvin, and Q is the electron charge ($1.60217646 \times 10^{-19}$ C) [21].

The solar PV system configuration is shown in Figure 12.3b. The basic power equation with solar PV voltage (V) and current (I) can be readily used to generate the IC algorithm as shown below [22]:

$$P = V \times I \qquad (12.11)$$

By differentiation of the aforementioned equation, we obtain [22]:

$$\frac{dP}{dV} = I + V \times \left(\frac{dI}{dV} \right) \tag{12.12}$$

The inclination dP/dV must equal zero in order for MPPT to function [22] as follows:

$$\frac{dI}{dV} = -\left(\frac{I}{V} \right) \tag{12.13}$$

12.2.3 USING CONTROL STRUCTURE IN ESS MODELLING

Because RERs have variable load requirements and intermittent power sources, ESS is essential for bridging the gap between power production and usage. Terminal voltage and SOC are two important characteristics to represent an ESS, as shown in the following example [23]:

$$V_{Bat} = V_{out} - i_{bat} R_{bat} - K \frac{Q}{Q - \int i_{bat}\, dt} + A e^{(-B \int i_b\, dt)} \tag{12.14}$$

$$SOC = 100 \left(1 - \frac{\int i_{bat}\, dt}{Q} \right) \tag{12.15}$$

where Q stands for capacity, A for the exponential voltage, K for the polarisation voltage, A for the exponential voltage, V_{out} for the open-circuit voltage, and R_{bat} for the internal resistance.

A bidirectional DC-DC buck-boost converter can be used to control charging and discharging. The measured current is equal to the charging and discharging value of the current reference. PI control is informed of the error. A two-loop control technique must be used in such case if the battery is being used to maintain the DC bus voltage [24]. The inner loop's reference current input value is determined by the PI controller's output signal, which is configured to be the difference between the measurement and reference DC voltage error. For instance, the external voltage control generates a negative value of the reference current when the value is higher than the reference value. At that time, the internal current loop modifies the duty cycle to force current flow from the DC side towards the ESS side, kicking off the battery charging process. After that, when the battery stores the excess energy, the DC voltage drops to the reference level. It is noted that the ESS limitations must take into account the SOC limits and the charging and discharging rates:

$$\begin{cases} SOC_{min} < SOC(k) < SOC_{max} \\ P_{ch}(k) \le P_{ch\,max} \\ P_{dch}(k) \le P_{dch\,max} \end{cases} \tag{12.16}$$

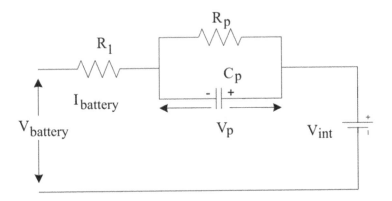

FIGURE 12.4 Battery model.

$$-\frac{P_{Bat}}{V_{Bat}} \le I_{Bat} \le \frac{P_{Bat}}{V_{Bat}} \tag{12.17}$$

where P_{Bat} and V_{Bat}, respectively, represent the battery's rated power and standard voltage. A battery model is shown in Figure 12.4, where V_{int} is the internal voltage and R_1 is the internal battery resistance, and R_p and C_p are the parasitic resistance and capacitance, respectively. The mathematical battery model's expression is as follows [22]:

$$V_{battery} = V_{int} + V_R \tag{12.18}$$

Figure 12.5a and b depict the battery ESS with buck-boost converter and the control block for the buck-boost converter, respectively. When the battery is being charged or discharged, the switches S_{B1} and S_{B2} are both complementarily ON. ESS-mode 1 (M1 with the switch S_{B1} ON and the switch S_{B2} OFF) and ESS-mode 2 (M2 with the switch S_{B1} OFF and the switch S_{B2} ON) are the two operational modes that are used. In mode 1, the converter operates as a grid-forming buck-boost converter, while the ESS regulates the DC bus voltage to the setting value. The reference input of the control strategy is a DC voltage V_{busref}. V_{bus} is the DC bus voltage, and V_b is the battery voltage. This enables mode 1 to implement the MG's independent operation. The inductor current serves as the control strategy's reference input value when the ESS is operating in mode 2 as a regulated current source. As a result, the ESS serves as a buck-boost converter that supports the grid. According to the command from the EMS, a selection of switching mechanism is utilised to alternate between the operational modes (modes 1 and 2) of the device.

12.2.4 USING A CONTROL STRUCTURE FOR IPC MODELLING

The AC filter's mathematical modelling can be expressed as:

$$V_i = V_g + R_f I_f + L_f \frac{dI_f}{dt} \tag{12.19}$$

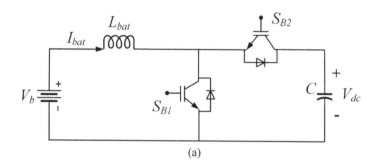

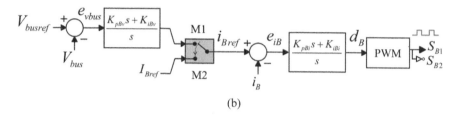

FIGURE 12.5   (a) Energy-storing battery system with a buck-boost converter. (b) For the buck-boost converter, the control block.

where I_f and L_f are the current and inductance of the filter, respectively, and R_f is the equivalent resistance of the filter and the lead wires. V_i and V_g are the voltage vectors of the AC terminal of the IPC and the grid bus, respectively. The LC filter capacitor's dynamic modelling can be expressed as:

$$C_f \frac{dV_c}{dt} = I_f - I_L \tag{12.20}$$

where C_f displays the filter capacitance, I_L displays the load current, and V_c indicates the capacitor voltage (in this example, $V_c = V_g$). The following is a representation of how these equations' continuous state-space modelling looks:

$$\frac{dx}{dt} = Ax + By \tag{12.21}$$

where

$$x = \begin{bmatrix} I_f \\ V_c \end{bmatrix} \tag{12.22}$$

$$y = \begin{bmatrix} V_i \\ I_L \end{bmatrix} \tag{12.23}$$

$$A = \begin{bmatrix} -R_f/L_f & -1/L_f \\ 1/C_f & 0 \end{bmatrix} \tag{12.24}$$

$$B = \begin{bmatrix} \frac{1}{L_f} \\ -\frac{1}{C_f} \end{bmatrix}$$ (12.25)

The battery can maintain a constant DC voltage when operating independently. In order to provide a controlled AC voltage, the IPC also functions as a VSI. The capacitor voltage must be the control target for this goal. A discrete state-space model is described as follows to determine the capacitor voltage at the following sampling period:

$$x(k+1) = A_d \cdot x(k) + B_d \cdot y(k)$$ (12.26)

where

$$A_d = e^{A.T_s}$$ (12.27)

$$B_d = \int_0^{T_s} e^{A.\tau} . B\, d\tau$$ (12.28)

where T_s represents the sampling period. For each potential input voltage, the filter voltages and currents are predicted using mathematical modelling. The most advantageous input voltage is chosen using a cost function (CF). The CF is defined as follows to manage the capacitor voltage during model predictive voltage control (MPVC) for stand-alone operation:

$$g_v = \sum_{j=1}^{n} \left[(V_{c\alpha}^{ref} - V_{c\alpha}^{k+j})^2 + (V_{c\beta}^{ref} - V_{c\beta}^{k+j})^2 \right]$$ (12.29)

where j stands for the prediction step, and $V_{c\alpha}$ and $V_{c\beta}$ denote the real and imaginary portions of the capacitor voltage in the reference frame, respectively. The vector that generates the smallest estimation of g_v in light of this cost objective will be chosen as the ideal value during the following sampling moment. Because the and parts are tightly managed, the Vc can adhere to its reference value. As a result, voltage stability with a sinusoidal waveform is possible. PLL is used to obtain the grid voltage/ frequency for grid transition. The reference voltage for the aforementioned CF is then set to a voltage with the same magnitude and a slightly lower amount of frequency. The AC voltage phasors of the MG and grid will rotate in this way at various rates while maintaining the same magnitudes. The mode transition will be established when the MG's voltage phase angle and the grid voltage are aligned. The IPC's output powers (P and Q) are

$$P = \mathrm{Re}\left\{ V_g \bar{I}_f \right\} = \frac{3}{2} (V_{g\alpha} I_{f\alpha} + V_{g\beta} I_{f\beta})$$ (12.30)

$$Q = \text{Im}\left\{V_g \bar{I}_f\right\} = \frac{3}{2}\left(V_{g\beta}I_{f\alpha} - V_{g\alpha}I_{f\beta}\right) \tag{12.31}$$

where the symbols '−' stand for complex conjugate operation, Re and Im represent real and imaginary values, $V_{g\alpha}$ and $V_{g\beta}$ are, respectively, grid voltages in the α and β reference frames, and $I_{f\alpha}$ and $I_{f\beta}$ are, respectively, filter voltages in the α and β reference frames. MPC has been frequently used to manage bidirectional converters due to advantages like flexible incorporation of control objectives and quick dynamic reaction [25,26]. Building up an MPC-based integrated EMS scheme with various operating modes of the IPC for MG applications is the key advancement. MPVC for stand-alone operation, grid transition operation, and model predictive power control (MPPC) in grid-tied (for future grid extension) are explicitly formulated. The IPC is used in grid-tied operation to maintain a constant DC voltage and to transfer power between the MG and the utility. The external loop, where the PI control is used to regulate the DC voltage, is where the reference P is acquired from. The DC voltage is adjusted to maintain the power balance within the network when resource conditions or consumption constraints change. Network modelling is used to anticipate the powers for the upcoming instant after getting the references. The best vector for the power converter is then selected using a CF as the criterion. The vector with the lowest value is ultimately chosen and used for the subsequent sampling instant. The prediction of the powers can be written as to calculate the instantaneous powers (active and reactive) in MPPC during grid-tied operation [27]:

$$\frac{d}{dt}\begin{bmatrix} P_i \\ Q_i \end{bmatrix} = -\frac{R_f}{L_f}\begin{bmatrix} P_i \\ Q_i \end{bmatrix} + \omega\begin{bmatrix} -Q_i \\ P_i \end{bmatrix} + \frac{3}{2L_f}\begin{bmatrix} \left(\left|v_c\right|^2 - \text{Re}\left(v_c\bar{v}_i\right)\right) \\ -\text{Im}\left(v_c\bar{v}_i\right) \end{bmatrix} \tag{12.32}$$

where a typical space vector is represented by $\bar{v}_i$ in $\alpha\beta$ reference with:

$$\begin{bmatrix} v_{i\alpha} \\ v_{i\beta} \end{bmatrix} = \frac{2}{3}V_{dc}\begin{bmatrix} S_{ia} - \frac{1}{2}(S_{ib} + S_{ic}) \\ \frac{\sqrt{3}}{2}(S_{ib} - S_{ic}) \end{bmatrix} \tag{12.33}$$

where S_{ia}, S_{ib}, S_{ic} denote the IPC's switching states. The power forecast for the following time is as follows for constant DC voltage:

$$\begin{bmatrix} P_i^{k+1} \\ Q_i^{k+1} \end{bmatrix} = T_s\left(-\frac{R_f}{L_f}\begin{bmatrix} P_i^k \\ Q_i^k \end{bmatrix} + \omega\begin{bmatrix} -Q_i^k \\ P_i^k \end{bmatrix} + \frac{3}{2L_f}\begin{bmatrix} \left(\left|v_c\right|^2 - \text{Re}\left(v_c\bar{v}_i\right)\right) \\ -\text{Im}\left(v_c\bar{v}_i\right) \end{bmatrix}\right) + \begin{bmatrix} P_i^k \\ Q_i^k \end{bmatrix} \tag{12.34}$$

where ω stands for grid frequency (in *rad*). Since the powers are the controller's primary concern when operating on a grid (P and Q). As a result, it is possible to evaluate the effects of each voltage vector on both powers and choose the vector that minimises the CF, which is described as follows:

$$g_{vsc} = \sqrt{\left(P^* - P_i^{k+1}\right)^2 + \left(Q^* - Q_i^{k+1}\right)^2} \tag{12.35}$$

where $k + 1$ is the subsequent sampling interval, and P^* and Q^* represent reference values for active and reactive powers, respectively.

12.3   IPC CONTROL METHOD BASED ON FCS-MPC

IPC serves as a connection between the MG's DC bus and AC bus and is a key component in the power balancing of the MG under study's DC and AC subsystems. As a result, the IPC's dynamic reaction performance is important for how the MG functions. Because of this, this section carefully examines the IPC's control issue.

12.3.1   IPC CONTROL METHODS BASED ON THE STUDIED MG'S OPERATION MODES

The bidirectional IPC (BIPC) of the MG is controlled using an FCS-MPC-based control system in order to regulate the most affected variable in terms of voltage or current under all conditions, including linear and non-linear modes. The proposed technique enables the BIPC to conduct either AC (inverter) or DC (rectification) operation in voltage control mode (VCM) or current control mode (CCM). Both the controlled currents and the controlled voltages are always effectively controlled during MG operation in both the VCM and the CCM. For instance, the MG's stand-alone operation mode controls the AC bus voltage. Similarly, in the MG's grid-connected operation mode, the AC current of the IPC is similarly controlled. Additionally, the IPC is necessary for the MG's operating mode switching in order to quickly control the transient process and maintain the stability of the MG system. The CF is defined as follows to satisfy control needs under various operation modes:

$$CF = \lambda_{dcCCM}(I_{dc}^* - I_{dc})^2 + \lambda_{dcVCM}(V_{dc}^* - V_{dc})^2$$
$$+ \lambda_{acCCM}(I_{ac}^* - I_{ac})^2 + \lambda_{acVCM}(V_{ac}^* - V_{ac})^2 \tag{12.36}$$

where the weight function for various modes, including CCM and VCM for corresponding AC and DC currents and voltages, is represented by λ. For matching DC and AC currents and voltages, the notation '*' indicates reference values; actual values are shown without this sign. The deviation of the most significant word will always have the largest weight value, meaning that the related weight will always be of maximum value. During this mode of operation, the remaining weights will remain zero.

1. IPC Operating in MG's Grid-Connected Operation Mode as an AC/DC Rectifier

The continuous state-space model of VSC is described below in relation to the creation of the active front-end (AFE) rectifier model, which is depicted in Figure 12.6:

$$L_s \frac{di_s}{dt} = v_s - v_r - R_s i_s \tag{12.37}$$

where L_s and R_s stand for the respective filter resistance and inductance. Source voltage, rectifier voltage, and input current are each denoted by the words v_s, v_r, and i_s. The discrete model is expressed as follows:

$$i_s(k+1) = \left(1 - \frac{R_s T_s}{L_s}\right) i_s(k) + \frac{T_s}{L_s}\left[v_s(k) - v_r(k)\right] \tag{12.38}$$

$$\begin{bmatrix} i_{s\alpha}(k+1) \\ i_{s\beta}(k+1) \end{bmatrix} = \left(1 - \frac{R_s T_s}{L_s}\right)\begin{bmatrix} i_{s\alpha}(k) \\ i_{s\beta}(k) \end{bmatrix} + \frac{T_s}{L_s}\begin{bmatrix} v_{s\alpha}(k) - v_{r\alpha}(k) \\ v_{s\beta}(k) - v_{r\beta}(k) \end{bmatrix} \tag{12.39}$$

This shows that the projected values of instantaneous powers (P and Q) and sampling time T_s are as follows:

$$P_i(k+1) = \mathrm{Re}\{v_s(k+1)i_s(k+1)\} = v_{s\alpha}i_{s\alpha} + v_{s\beta}i_{s\beta} \tag{12.40}$$

$$Q_i(k+1) = \mathrm{Im}\{v_s(k+1)i_s(k+1)\} = v_{s\beta}i_{s\alpha} - v_{s\alpha}i_{s\beta} \tag{12.41}$$

The following examples show how the CF is used for the predictive power control:

$$g_{AFE} = |Q_i^* - Q_i(k+1)| + |P_i^* - P_i(k+1)| = g_r \tag{12.42}$$

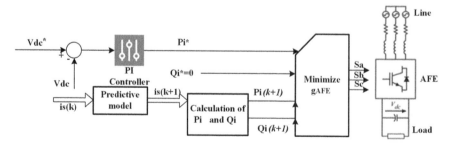

FIGURE 12.6   Block diagram for the MPPC used in the IPC as AFE.

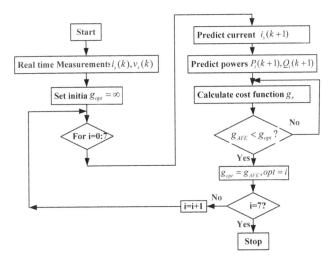

FIGURE 12.7   Flowchart of the MPPC algorithm used in the IPC as AFE.

where Q_i^* and P_i^* stand for P and Q's respective references. The flow diagram of the FCS-MPC algorithm used to control the AFE functioning of the IPC is shown in Figure 12.7.

2. IPC Operating in MG's Stand-Alone Operation Mode as a DC/AC Inverter
   The following mathematical equation describes the MPVC:

$$\frac{d}{dt}\begin{bmatrix} i_f \\ v_c \\ i_L \end{bmatrix} = \begin{bmatrix} 0 & -\dfrac{1}{L} & 0 \\ \dfrac{1}{C} & 0 & -\dfrac{1}{C} \\ 0 & 0 & 0 \end{bmatrix}\begin{bmatrix} i_f \\ v_c \\ i_L \end{bmatrix} + \begin{bmatrix} \dfrac{1}{L} \\ 0 \\ 0 \end{bmatrix} v_i + \begin{bmatrix} 0 \\ 0 \\ f(i_L, v_c) \end{bmatrix} \tag{12.43}$$

where respectively, V_c, V_i, I_f, R_f, and L stand for the load voltage, the invert voltage vectors, the filter current, the filter resistance, and the filter inductance. The relationship described above can be modelled in state space as follows:

$$\frac{dx}{dt} = Ax + Bv_i + u_d(k) \tag{12.44}$$

The discrete mathematical model is written as follows in order to determine the voltage vector for the subsequent sampling period:

$$x(k+1) = A_q x(k) + B_q v_i(k) + u_d(k) \tag{12.45}$$

The following relationship can be obtained by resolving the aforementioned relationship for the discrete state space:

$$\begin{bmatrix} V_c \\ I_f \end{bmatrix}^{k+1} = A_d \begin{bmatrix} V_c \\ I_f \end{bmatrix}^{k} + B_d \begin{bmatrix} V_i \\ I_L \end{bmatrix} \tag{12.46}$$

where

$$A_q = e^{AT_s} \tag{12.47}$$

$$B_q = \int_0^{T_s} e^{A\tau} B d\tau \tag{12.48}$$

For VSC switches (i.e., S_a, S_b, and S_c), V_i represents the voltage vector with seven switching states (i.e., $i=0$ and $i=7$ has the same value) with the relation:

$$V_i = \begin{cases} \dfrac{2}{3} V_{dc} e^{j(i-1)\frac{\pi}{3}} & for \ \ i=1,2,...6 \\ 0 & for \ \ i=0,7 \end{cases} \tag{12.49}$$

The following objective function is used when using the external disturbance observer:

$$g_{inv} = (v_{c\alpha}^* - v_{c\alpha})^2 + (v_{c\beta}^* - v_{c\beta})^2 \tag{12.50}$$

The algorithm for IPC in an inverter mode with solo operation is shown in Figure 12.8. The output frequency controls the actual power (P), whereas the voltage of the VSC controls the imaginary power (Q). To control the power, the output voltage and frequency of VSC are used (P and Q).

$$\omega_i = \omega_{nom} - m_i P_i \tag{12.51}$$

$$V_i = V_{nom} - n_i Q_i \tag{12.52}$$

With the following relations, the coefficients (m and n) are chosen to guarantee a steady operation:

$$m_i = \frac{\Delta w}{Q_{max}} \tag{12.53}$$

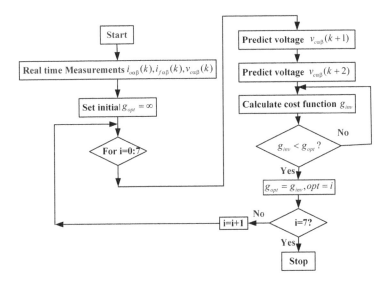

FIGURE 12.8   Flowchart of the MPVC algorithm used in the IPC.

$$n_i = \frac{\Delta V}{P_{\text{max}}} \tag{12.54}$$

where there are the IPC's maximum power restrictions. The following terms Δw and ΔV display, respectively, the frequency/voltage deviations with an upper limit of the IPC output. The values chosen for (m) and (n) are 0.00140 and 0.00080, respectively.

12.4   FOR THE RESEARCHED MG, AN INTEGRATED EMS WITH COORDINATED CONTROL

To ensure dependable power delivery for an MG, multiple operation modes must be taken into account. According to the power balancing equation below, the power inside the MG needs to be adjusted:

$$\begin{cases} \sum_{j=1}^{N} P_{PV} + \sum_{j=1}^{N} P_{WT} + P_g + P_{bat} = P_{load} \\ \\ \sum_{j=1}^{N} P_{PV} + \sum_{j=1}^{N} P_{WT} - P_{load} = P_{net} \end{cases} \tag{12.55}$$

where the power that the MG exchanges with the utility is represented by P_g, where positive P_g denotes taking power from the utility, and negative P_g denotes supplying to the utility. P_{bat} stands for the output power of the battery ESS, while P_{bat} is positive when discharging and negative when charging. P_{net} can be defined as P_{PV} and P_{WT} less P_{Load}. Positive P_{net} denotes the presence of excess renewable energy produced by the

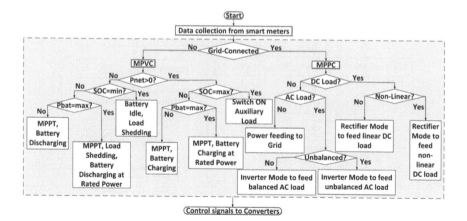

FIGURE 12.9 Diagram of the suggested integrated EMS algorithm for the understudied MG.

MG, some of which must be delivered to the utility or used to charge a battery ESS; a portion of the load must now use electricity from the grid, a battery ESS, or both since negative P_{net} implies that there is insufficient power coming from the distributed renewable energy generation systems in the MG.

An MG's EMS must be integrated in order to meet the criteria for cheap cost, and it should effectively integrate coordinated control for the MG's DGs, ESS, and IPC. The creation of an integrated EMS with coordinated control for the reliable and effective operation of the MG under study in this chapter is illustrated in this section. Figure 12.9 illustrates the algorithm branches of the integrated EMS proposal. The algorithm of the suggested integrated EMS for MG has several branches. Grid connectivity, SOC, P_{net}, and other factors are taken into consideration while choosing the algorithm's subsequent branching. The correspondences between the algorithm's branches and the operation modes of the MG under study are explained in the next section using three subsections.

12.4.1 MODE 1 OPERATION (STANDALONE)

In this instance, MG's independent operation is discussed. The MG and the main grid don't exchange any power. The power needs to be changed inside the MG because it becomes an island system. The ESS uses a two-loop (inner current and outer voltage) control to maintain a steady DC voltage during this islanded operation. Using the MPVC approach, the IPC regulates and manages the AC voltage. The rate of battery charging is adjusted during the stand-alone time in accordance with the priorities established by the integrated EMS, and ESS is successful in suppressing the transients and/or fluctuations brought on by the PV/WT power generation and/or load variations. As a result, ESS is successfully managing the DC bus variations through DC bus voltage control. Three unique circumstances should be taken into account:

1. *Higher demand and lower wind/irradiance:* In this instance, the load demand is not being met by the PV/WT power output. In order to accommodate the additional load demand while the ESS is in discharging mode, both the PV and WT must run in MPPT modes. When the ESS's SOC falls below a certain threshold or its power discharge exceeds the battery's capacity, a load-shedding scenario is initiated to ensure that the most vital loads are always supplied with power.

2. *Lower demand and higher wind/irradiance:* When taking this into account, the PV/WT is producing more power than the necessary load requirement. As a result, the ESS is absorbing additional energy during charging. The MPPT operating modes of the PV and wind are still in use, but the auxiliary load will be turned ON to service the extra generated power if the ESS stored energy exceeds its rating or the ESS reaches its maximum limit of SOC.

3. *Power exchange:* The AC and DC buses are coordinated with power exchange in this scenario using interconnected power converters (IPCs). The IPC functions as an inverter to one component and as a rectifier to the subsequent portion for the unidirectional power flow between two AC/DC MG sections [25]. The exchange is essential for ensuring proper power balance between the two segments and maintaining the voltages within allowable ranges [26].

12.5   RESULTS AND DISCUSSION OF THE SIMULATION

The electrical parameters of the suggested model, along with the proportional integral (PI) controller parameters with WT and ESS, are provided in Table 12.1. The following analysis looks at the investigations for the offered techniques.

TABLE 12.1A
Parameters for Electrical Simulation

Parameter	Symbol	Value
PV specifications	–	SunPower SPR-305
PV power (upper limit)	–	16.50 kW
WT power (max)	–	4.000 kW
RLC filter resistance	R_f	0.100 Ω
RLC filter inductance	L_f	10.00 mH
RLC filter capacitance	C_f	250.0 μF
Load resistance	R_L	25.00 Ω (variable)

TABLE 12.1B
Electrical Parameters for Simulation

Frequency	f	50.00 Hz
DC voltage	Vdc	510.0 V
Grid voltage	$V_{g, RMS}$	110.0 V
DC link capacitor	C	2,000 μF
Sampling time	T_s	33.00 μs
PI controller (WT)	K_p, K_i	0.0510, 1.0000
PI controller (battery inner loop)	K_p, K_i	4.9830, 10.7520
PI controller (outer battery loop)	K_p, K_i	0.4490, 1.0650

12.5.1   STEP CHANGE IN AC LOAD WITH HIGHER SOC IN ISLANDED MODE

When the MPC technique is used with a larger SOC limit, the impact of external generation/load disruptions is seen. As shown in Figure 12.10a, the waveforms of voltage and current are obtained with a smooth contour. The graphic shows that the load switch-2 is turned ON to inject the secondary load power as soon as the SOC limit hits the upper limit or 80% (Figure 12.10b). Power sharing is depicted in Figure 12.10c as the load requirement is met. The ESS receives the additional power while it is charging, and it also handles sudden changes.

12.5.2   STEP CHANGE IN AC LOAD WITH LOWER SOC IN ISLANDED MODE

When using the MPC method with the lower SOC limit, the impact of outside disturbances is also examined. As shown in Figure 12.11a, the waveforms of voltage and current are obtained with a smooth contour. The diagram shows that when the SOC drops to 20%, the switch-1 is turned on and the next primary load is connected, ensuring that the extra power is used by the load (Figure 12.11b). The power sharing across various MG components while meeting the necessary load requirement is shown in Figure 12.11c. During the charging phase, the ESS absorbs the extra power while the ESS's stored energy manages unexpected shifts.

12.5.3   REGULATION OF DC VOLTAGE IN RECTIFICATION
MODE WITH CONSTANT DC LOAD

Figure 12.12 illustrates the implementation of a fixed DC load for the AFE procedure. The 510 V reference value of DC voltage is tracked by the MPC algorithm. Additionally, the constant AC voltage is attained. It's interesting to note that the rise time for the actual DC voltage to follow the reference DC voltage through the MPC algorithm causes the AC source current to exhibit strong transients during the first 30 ms. While the DC load is receiving 5.3 kW of power from the P, the power Q is set to zero ($Q = 0$).

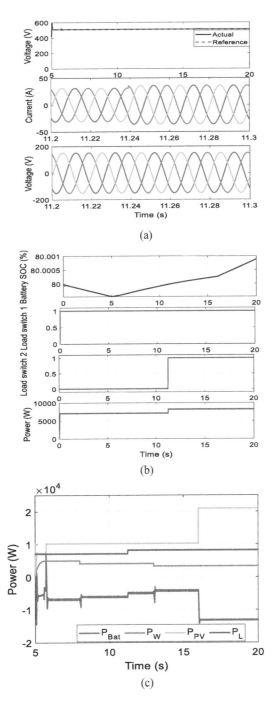

FIGURE 12.10 MPC with (a) DC/AC voltage and current. (b) Higher SOC, load switches, and power. (c) Power sharing.

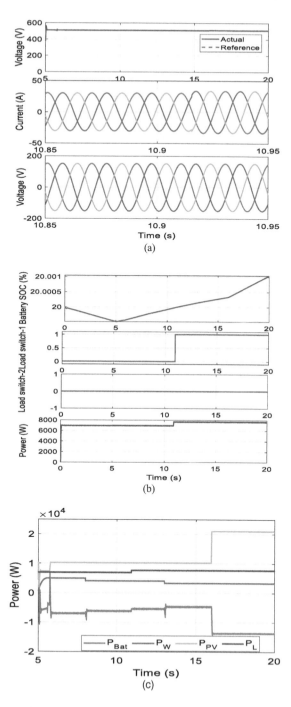

FIGURE 12.11 MPC with (a) Current, DC/AC voltages. (b) A reduced SOC, load switches, and power. (c) Power sharing.

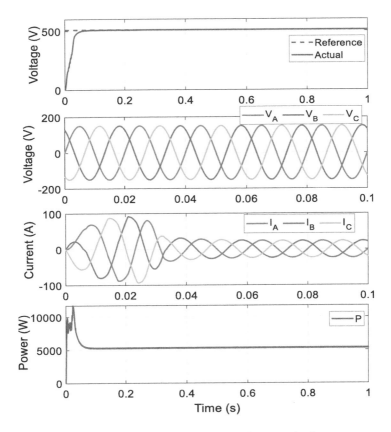

FIGURE 12.12   AFE rectification under DC voltage and constant load.

12.5.4   UNDER DC VOLTAGE VARIATION, DC VOLTAGE REGULATION DURING RECTIFICATION

To evaluate the effectiveness of the suggested MPC method, a DC voltage is changed from 510 V to 400 V at 0.5 s in the subsequence operation (Figure 12.13). The graph shows how precisely the controller is following the reference voltage. The three-phase current was reduced to 37.5% while the active power P was cut to 38.7% by a 27.5% decrease in DC voltage (3.25 kW). There was no change in the reactive power Q.

12.5.5   UNDER LOW DC LOAD VARIATION, RECTIFICATION MODE DC VOLTAGE REGULATION

To observe the DC voltage, the DC load is changed between 50 and 75 Ω at 0.5 s (Figure 12.14).

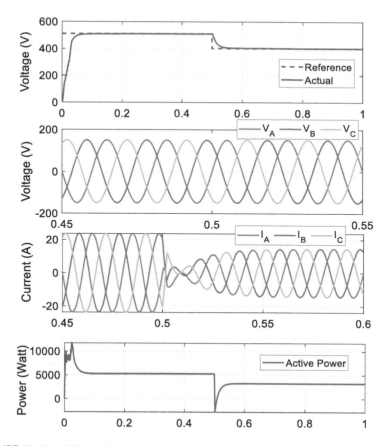

FIGURE 12.13 AFE rectification for DC voltage and changing load.

12.5.6 UNDER HIGH DC LOAD VARIATION, RECTIFICATION MODE AC VOLTAGE REGULATION

The controller is watching the DC reference value of the voltage following the unexpected load fluctuation, and a 3.5% variance in the DC voltage (i.e., 328 V) is seen during the sudden load fluctuation (Figure 12.15). While P is adjusted between 2.0 and 3.5 kW, a 33.3% drop in three-phase current is seen. The DC voltage is kept constant as the DC load's real power changes from 2 to 3.5 kW.

12.5.7 REGULATING THE AC VOLTAGE WHILE USING AN ISLANDED INVERTER WITH A CONSTANT AC LOAD

The stand-alone MG design is used in the following analysis to regulate the voltage during grid-forming operation, and voltage MPC (MPVC) is used. At a constant DC voltage of 510 V, the balanced and fixed load is used. In Figure 12.16, the load current and regulated AC voltage are shown, and 3 kW of AC power is used.

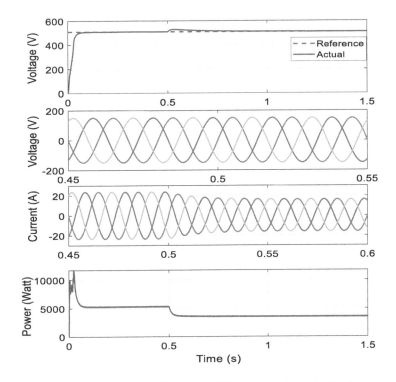

FIGURE 12.14   AFE rectification with constant DC voltage and varying load.

12.5.8   REGULATION OF AC VOLTAGE IN ISLANDED INVERTER MODE WITH VARIABLE AC LOAD

The AC load varies from 20 to 10 Ω at 0.5 s. At 510 V, the DC source voltage remains constant. While the Q is zero ($Q=0$), the P is changed between 1.69 and 3.37 kW at 0.5 s. The load voltage, which is managed by the MPC, is shown in Figure 12.17. Between 7.5A and 15A, three-phase current is used, with a 0.5 s load fluctuation.

12.5.9   UNDER A NON-LINEAR AC LOAD, AN ISLANDED INVERTER IS USED

With the modification of P and Q, the linear load of 20 Ω is switched out for a non-linear load of 25 Ω at 0.5 s (Figure 12.18). It is clear that the suggested MPC-based approach ensures an efficient voltage profile even under non-linear loads.

12.5.10   COMPARISON OF LOAD RIPPLES AND THD

Inverter mode THD is displayed in Table 12.2. Table 12.3 displays THD for various IPC modes. Table 12.4 displays the THD for the voltage and current waveforms using MPC and other control methods. The outcomes confirmed the suggested MPC-based rural system's superior performance. It is possible to get THD of 0.28% with MPC and 3.71% with PI management.

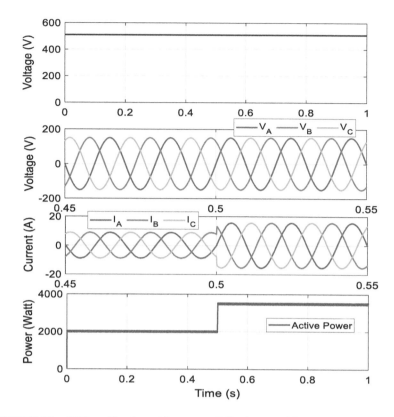

FIGURE 12.15 AFE rectification with constant DC voltage and fluctuating high load.

The THD comparison in Table 12.5 is similar. The MPC's THD is 0.28%, which is lower than the referenced THD in [33] under comparable circumstances. The specific component sizing, DC/AC voltage level, and load profile may be the cause. Another WT-ESS model that is investigated in [32] exhibits nearly identical THD under the MPC without a PV source. With PI control, the load power swells are seen. Less THD and power swells with the MPC demonstrate its efficient operation. THD and power quality in inverter mode are shown in Table 12.6.

When PV and WT generation is intermittent and there is fluctuating load demand, MPC-based integrated EMS performs well. High power quality and voltage are attained for an islanded/grid-connected MG with the use of the suggested MPC. The implemented MPC is durable when subjected to transient analysis and fluctuating DC voltage levels. According to IEEE standards 929, 2,000, and 519, just 0.28% THD of the AC output voltage is produced, which is significantly less than 5% [34].

12.6 CONCLUSION

For rural MGs, an integrated EMS with a pure green energy system is presented. In this system, IPC control is designed using MPC. In Pakistan, under various

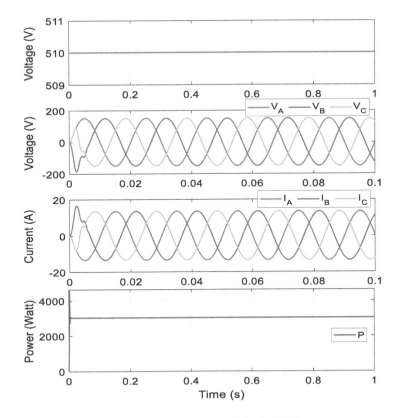

FIGURE 12.16 With constant load/DC voltage and islanded VSI.

conditions of priority load requirements and the availability of RERs, integrated EMS also works in coordination with other conventional controllers that are developed for practical distributed energy resources (DERs) in rural MGs. The tested EMS shows positive performance under a variety of load variations, varying WT speeds, and varying PV irradiance circumstances. Coordinated control based on FCS-MPC is used for a PV-WT-ESS system with better output power quality. In the case of variable DC bus voltage, the suggested MPC-based EMS technique requires less rise and fall time than the traditional PI control scheme. With no steady-state error and stable rural MG operation, the proposed methodology enhanced the voltage and completely tracked the reference voltage. For the load voltage, THD of 0.28% is attained. The proposed MPC approach ensured higher power quality in steady state and transient response of the system by controlling voltage magnitude. Additionally explained by the changing power supply and sudden load fluctuations is EMS with battery SOC. To maximise WT power, solar PV power, and DC bus voltage regulation, PI controllers are employed.

As technical, economic, and environmental issues are taken into account at the same time, the EMS and optimisation control in MG systems are evolving into multi-objective (management and optimisation) functions. As a result, several techniques,

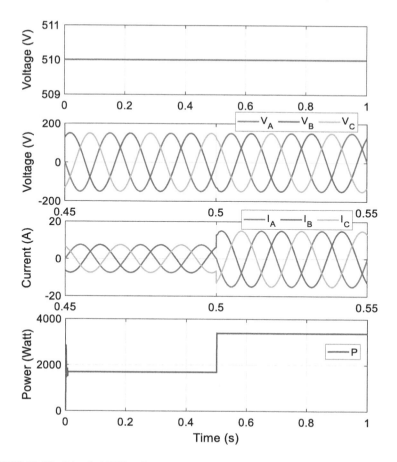

FIGURE 12.17   Islanded VSI under constant DC voltage and varying load.

including precise, stochastic, and predictive ones, have been developed for EMS. These techniques were chosen for the MG setting based on their applicability, dependability, and resource availability. Predictive control, on the other hand, is a strong contender because it combines optimum control with multivariable processes and is a more adaptable control method, making it simple to include MG system limitations and optimisation functions. Power-smoothing applications and uncertainty are no match for predictive control. As a result, it is simple to develop multiobjective control and constraint functions for the same control scheme. Additionally, the implementation of predictive control in real-world scenarios necessitates a comprehensive platform that combines MG elements with all machinery for measuring and forecasting significant input data sets. Predictive control can be found to be a promising replacement for EMS in MG systems, with particular application in MGs, thanks to current technological advancements in microprocessors, data analysis, and machine learning.

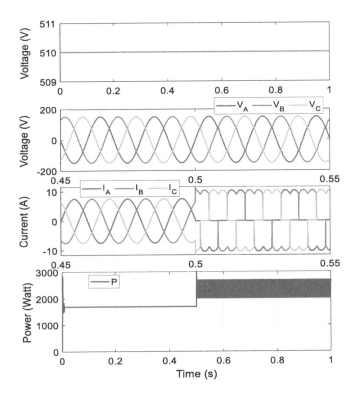

FIGURE 12.18   VSI with an island and a non-linear load.

TABLE 12.2
THD throughout MPC and PI (Inverter Operation)

Load	PI	FCS-MPC
No load	1.220	0.570
Resistive (balanced)	1.220	0.570
Non-linear	4.310	1.000

TABLE 12.3
THD Using the MPC Approach

VSC Mode	THD	
	Current	Voltage
AFE rectifier	0.560	–
Grid-connected	0.430	–
Islanded (linear load)	0.560	0.500
Islanded (non-linear load)	–	1.000

TABLE 12.4
THD Using Several Control Methods (Inverter Mode)

Ref.	Control Strategy	THD (voltage)	
		Non-linear load	Linear load
[27]	PR	4.60	1.40
[28]	SMC	2.66	–
[29]	PI	42.0	16.0
[30]	Deadbeat	4.80	2.10
Current study	MPC	1.0	0.50

TABLE 12.5
Comparison of PI and Implemented MPC Techniques

Ref	Studied Model	Voltage Level	Voltage Magnitude Error (%)	% THD (V)	% THD (I)
[31]	PV-WT-BSS	326.70	−5.048	1.830	2.670
[32]	WT-BSS	308.70	0.739	0.260	–
[33]	PV-WT-BSS	314.90	−1.254	0.300	0.300
Present	PV-WT-BSS	151.90	−1.266	0.280	0.280

TABLE 12.6
An Evaluation of Power Quality and THD in Comparison (Inverter Mode)

Ref/System	DC/AC (V)	Sizes	Load	Method (%THD)	Load Ripples
[35]/WT-ESS	710/311	PV: 0	11.25	PI (3.71)	±1,000 W
		WT: 2.4		MPC (0.28)	±20 W
		BSS: 53.9			
[36]/PV-WT-ESS	750/311	PV: 13.4	21.57	PI (5.44)	±480W
		WT: 4.0		MPC (0.30)	±30 W
		BSS: 47.4			
Present/ PV-WT-ESS	510/150	PV: 21.1	35.94	PI (3.71)	±1,500 W
		WT: 5.0		MPC (0.28)	±30 W
		BSS: 90.1			

NOMENCLATURE

AFE   Active front end
BIPC   Bidirectional IPC
CCM   Current control mode
CF   Cost function
DERs   Distributed energy resources
DGs   Distributed generations

EMS Energy management system
ESS Energy storage system
FCS Finite control set
IC Incremental conductance
IPC Interlinking power converter
MG Microgrid
MPC Model predictive control
MPP Maximum power point
MPPC Model predictive power control
MPPT Maximum power point tracking
MPVC Model predictive voltage control
PCC Point of common coupling
PMSG Permanent magnet synchronous generator
PV Photovoltaic
P-V Power-voltage
PWM Pulse width modulation
RERs Renewable energy resources
RLC Resistor-Inductor-Capacitor
SOC State of charge
THD Total harmonic distortion
VCM Voltage control mode
VSI Voltage source inverter
WT Wind turbine

REFERENCES

[1] J. Liu, Y. Miura, and T. Ise, "Cost-Function-Based Microgrid Decentralized Control of Unbalance and Harmonics for Simultaneous Bus Voltage Compensation and Current Sharing," *IEEE Transactions on Power Electronics*, vol. 34, no. 8, pp. 7397–7410, 2019, doi: 10.1109/TPEL.2018.2879340.

[2] Y. Han, H. Li, P. Shen, E. A. A. Coelho, and J. M. Guerrero, "Review of Active and Reactive Power Sharing Strategies in Hierarchical Controlled Microgrids," *IEEE Transactions on Power Electronics*, vol. 32, no. 3, pp. 2427–2451, 2017, doi: 10.1109/TPEL.2016.2569597.

[3] J. Fang, G. Xiao, X. Yang, and Y. Tang, "Parameter Design of a Novel Series-Parallel-Resonant LCL Filter for Single-Phase Half-Bridge Active Power Filters," *IEEE Transactions on Power Electronics*, vol. 32, no. 1, pp. 200–217, 2017, doi: 10.1109/TPEL.2016.2532961.

[4] J. Kukkola and M. Hinkkanen, "State Observer for Grid-Voltage Sensorless Control of a Converter Equipped with an LCL Filter: Direct Discrete-Time Design," *IEEE Transactions on Industry Applications*, vol. 52, no. 4, pp. 3133–3145, 2016, doi: 10.1109/TIA.2016.2542060.

[5] M. Ben Said-Romdhane, M. W. Naouar, I. Slama-Belkhodja, and E. Monmasson, "Robust Active Damping Methods for LCL Filter-Based Grid-Connected Converters," *IEEE Transactions on Power Electronics*, vol. 32, no. 9, pp. 6739–6750, 2017, doi: 10.1109/TPEL.2016.2626290.

[6] W. Lei, W. Jiang, Y. Wang, and H. Huang, "Different Control Objectives Under Stationary frame for Grid-Connected Converter Under Unbalanced Grid Voltage," *2016 IEEE 8th International Power Electronics and Motion Control Conference, IPEMC-ECCE Asia 2016*, pp. 725–730, 2016, doi: 10.1109/IPEMC.2016.7512375.

[7] S. Xie, X. Wang, C. Qu, X. Wang, and J. Guo, "Impacts of Different Wind Speed Simulation Methods on Conditional Reliability Indices," *International Transactions on Electrical Energy Systems*, vol. 20, no. February, 2011, pp. 1–6, 2013, doi: 10.1002/etep.

[8] Y. Shan, J. Hu, M. Liu, J. Zhu, and J. M. Guerrero, "Model Predictive Voltage and Power Control of Islanded PV-Battery Microgrids with Washout Filter Based Power Sharing Strategy," *IEEE Transactions on Power Electronics*, pp. 1–1, 2019, doi: 10.1109/tpel.2019.2930182.

[9] Q. Wei, L. Xing, D. Xu, B. Wu, and N. R. Zargari, "Modulation Schemes for Medium-Voltage Current Source Converter-Based Drives: An Overview," *IEEE Journal of Emerging and Selected Topics in Power Electronics*, vol. 6777, no. c, 2018, doi: 10.1109/JESTPE.2018.2831695.

[10] X. Lu, K. Sun, J. M. Guerrero, J. C. Vasquez, and L. Huang, "State-of-Charge Balance Using Adaptive Droop Control for Distributed Energy Storage Systems in DC Microgrid Applications," *IEEE Transactions on Industrial Electronics*, vol. 61, no. 6, pp. 2804–2815, 2014, doi: 10.1109/TIE.2013.2279374.

[11] M. Jayachandran and G. Ravi, "Decentralized Model Predictive Hierarchical Control Strategy for Islanded AC Microgrids," *Electric Power Systems Research*, vol. 170, no. December, 2018, pp. 92–100, 2019, doi: 10.1016/j.epsr.2019.01.010.

[12] P. Acuna, L. Moran, M. Rivera, J. Dixon, and J. Rodriguez, "Improved Active Power Filter Performance for Renewable Power Generation Systems," *IEEE Transactions on Power Electronics*, vol. 29, no. 2, pp. 687–694, 2014, doi: 10.1109/TPEL.2013.2257854.

[13] S. Rehman, H. U. R. Habib, S. Wang, M. S. Buker, L. M. Alhems, and H. Z. Al Garni, "Optimal Design and Model Predictive Control of Standalone HRES: A Real Case Study for Residential Demand Side Management," *IEEE Access*, vol. 8, pp. 29767–29814, 2020, doi: 10.1109/ACCESS.2020.2972302.

[14] A. Khamis, H. M. Nguyen, and D. S. Naidu, "Nonlinear, Optimal Control of Wind Energy Conversion Systems Using Differential SDRE," *Proceedings -2015 Resilience Week, RSW 2015*, pp. 86–91, 2015, doi: 10.1109/RWEEK.2015.7287423.

[15] P. Kumar, R. Kumar, A. Verma, and M. C. Kala, "Simulation and Control of WECS with Permanent Magnet Synchronous Generator (PMSG)," *Proceedings -2016 8th International Conference on Computational Intelligence and Communication Networks, CICN 2016*, pp. 516–521, 2017, doi: 10.1109/CICN.2016.107.

[16] K. Tazi, M. F. Abbou, and F. Abdi, *Performance Analysis of Micro-Grid Designs with Local PMSG Wind Turbines*, no. 0123456789. Springer Berlin Heidelberg, 2019. doi: 10.1007/s12667-019-00334-2.

[17] A. Merabet, K. Tawfique Ahmed, H. Ibrahim, R. Beguenane, and A. M. Y. M. Ghias, "Energy Management and Control System for Laboratory Scale Microgrid Based Wind-PV-Battery," *IEEE Transactions on Sustainable Energy*, vol. 8, no. 1, pp. 145–154, Jan. 2017, doi: 10.1109/TSTE.2016.2587828.

[18] G. M. Masters, *Renewable and Efficient Electric Power Systems*. 2004. doi: 10.1002/0471668826.

[19] I. V. Banu, R. Beniuga, and M. Istrate, *Comparative Analysis of the Perturb-and-Observe and Incremental Conductance MPPT Methods*, 2013. doi: 10.1109/ATEE.2013.6563483.

[20] N. Priyadarshi, S. Padmanaban, P. Kiran Maroti, and A. Sharma, "An Extensive Practical Investigation of FPSO-Based MPPT for Grid Integrated PV System Under Variable Operating Conditions With Anti-Islanding Protection," *IEEE Systems Journal*, pp. 1–11, 2018, doi: 10.1109/JSYST.2018.2817584.

[21] O. Santollani, C. Ruiz, and D. Ruiz, "An Energy Management System," *Hydrocarbon Engineering*, vol. 13, no. 9, pp. 51–58, 2008.

[22] L. A. G. Gomez, A. P. Grilo, M. B. C. Salles, and A. J. S. Filho, "Combined Control of DFIG-Based Wind Turbine and Battery Energy Storage System for Frequency Response in Microgrids," *Energies*, vol. 13, no. 4, 2020, doi: 10.3390/en13040894.

[23] O. Tremblay, L.-A. Dessaint, and A.-I. Dekkiche, "A Generic Battery Model for the Dynamic Simulation of Hybrid Electric Vehicles," *2007 IEEE Vehicle Power and Propulsion Conference*, 2007, pp. 284–289. doi: 10.1109/VPPC.2007.4544139.

[24] T. Ma, M. H. Cintuglu, and O. A. Mohammed, "Control of a Hybrid AC/DC Microgrid Involving Energy Storage and Pulsed Loads," *IEEE Transactions on Industry Applications*, vol. 53, no. 1, pp. 567–575, 2017, doi: 10.1109/TIA.2016.2613981.

[25] M. Shahidehpour, Z. Li, W. Gong, S. Bahramirad, and M. Lopata, "A Hybrid AC/DC Nanogrid: The Keating Hall Installation at the Illinois Institute of Technology," *IEEE Electrification Magazine*, vol. 5, no. 2, pp. 36–46, 2017, doi: 10.1109/MELE.2017.2685858.

[26] J. M. Guerrero, J. C. Vasquez, J. Matas, L. G. De Vicuña, and M. Castilla, "Hierarchical Control of Droop-Controlled AC and DC Microgrids - A General Approach Toward Standardization," *IEEE Transactions on Industrial Electronics*, vol. 58, no. 1, pp. 158–172, 2011, doi: 10.1109/TIE.2010.2066534.

[27] A. Hasanzadeh, O. C. Onar, H. Mokhtari and A. Khaligh, "A Proportional-Resonant Controller-Based Wireless Control Strategy with a Reduced Number of Sensors for Parallel-Operated UPSs," in *IEEE Transactions on Power Delivery*, vol. 25, no. 1, pp. 468–478, January 2010, doi: 10.1109/TPWRD.2009.2034911.

[28] H. Komurcugil, "Rotating-Sliding-Line-Based Sliding-Mode Control for Single-Phase UPS Inverters," *IEEE Transactions on Industrial Electronics*, 2012, doi: 10.1109/TIE.2011.2159354.

[29] P. C. Loh, M. J. Newman, D. N. Zmood, and D. G. Holmes, "A Comparative Analysis of Multiloop Voltage Regulation Strategies for Single and Three-Phase UPS Systems," *IEEE Transactions on Power Electronics*, 2003, doi: 10.1109/TPEL.2003.816199.

[30] P. Mattavelli, "An Improved Deadbeat Control for UPS Using Disturbance Observers," *IEEE Transactions on Industrial Electronics*, 2005, doi: 10.1109/TIE.2004.837912.

[31] P. S. Sikder and N. Pal, "Modeling of an Intelligent Battery Controller for Standalone Solar-Wind Hybrid Distributed Generation System," *Journal of King Saud University - Engineering Sciences*, no. xxxx, 2019, doi: 10.1016/j.jksues.2019.02.002.

[32] H. U. R. Habib, S. Wang, M. R. Elkadeem, and M. F. Elmorshedy, "Design Optimization and Model Predictive Control of a Standalone Hybrid Renewable Energy System: A Case Study on a Small Residential Load in Pakistan," *IEEE Access*, vol. 7, pp. 117369–117390, 2019, doi: 10.1109/access.2019.2936789.

[33] E. A. Al-Ammar et al., "Residential Community Load Management Based on Optimal Design of Standalone HRES with Model Predictive Control," *IEEE Access*, vol. 8, pp. 12542–12572, 2020, doi: 10.1109/ACCESS.2020.2965250.

[34] O. Krishan and S. Suhag, "Techno-Economic Analysis of a Hybrid Renewable Energy System for an Energy Poor Rural Community," *Journal of Energy Storage*, vol. 23, no. April, pp. 305–319, 2019, doi: 10.1016/j.est.2019.04.002.

[35] H. U. R. Habib, S. Wang, M. R. Elkadeem, and M. F. Elmorshedy, "Design Optimization and Model Predictive Control of a Standalone Hybrid Renewable Energy System: A Case Study on a Small Residential Load in Pakistan," *IEEE Access*, vol. PP, pp. 1–1, 2019, doi: 10.1109/ACCESS.2019.2936789.

[36] E. A. Al-ammar et al., "Based on Optimal Design of Standalone HRES with Model Predictive Control," *IEEE Access*, vol. PP, p. 1, 2020, doi: 10.1109/ACCESS.2020.2965250.

Index

Note: **Bold** page numbers refer to tables and *italic* page numbers refer to figures.